T R A N S I T   C O O P E R A T I V E   R E S E A R C H   P R O G R A M

Report 11

Impact of Radio Frequency Refarming on Transit Communications

JEFFREY J. MAUL and JOHN J. GREICHEN
Arthur D. Little, Inc.
Cambridge, MA

Subject Area

Public Transit

Research Sponsored by the Federal Transit Administration in Cooperation with the Transit Development Corporation

TRANSPORTATION RESEARCH BOARD
NATIONAL RESEARCH COUNCIL

NATIONAL ACADEMY PRESS
Washington, D.C. 1996

TRANSIT COOPERATIVE RESEARCH PROGRAM

The nation's growth and the need to meet mobility, environmental, and energy objectives place demands on public transit systems. Current systems, some of which are old and in need of upgrading, must expand service area, increase service frequency, and improve efficiency to serve these demands. Research is necessary to solve operating problems, to adapt appropriate new technologies from other industries, and to introduce innovations into the transit industry. The Transit Cooperative Research Program (TCRP) serves as one of the principal means by which the transit industry can develop innovative near-term solutions to meet demands placed on it.

The need for TCRP was originally identified in *TRB Special Report 213—Research for Public Transit: New Directions,* published in 1987 and based on a study sponsored by the Urban Mass Transportation Administration—now the Federal Transit Administration (FTA). A report by the American Public Transit Association (APTA), *Transportation 2000,* also recognized the need for local, problem-solving research. TCRP, modeled after the longstanding and successful National Cooperative Highway Research Program, undertakes research and other technical activities in response to the needs of transit service providers. The scope of TCRP includes a variety of transit research fields including planning, service configuration, equipment, facilities, operations, human resources, maintenance, policy, and administrative practices.

TCRP was established under FTA sponsorship in July 1992. Proposed by the U.S. Department of Transportation, TCRP was authorized as part of the Intermodal Surface Transportation Efficiency Act of 1991 (ISTEA). On May 13, 1992, a memorandum agreement outlining TCRP operating procedures was executed by the three cooperating organizations: FTA, the National Academy of Sciences, acting through the Transportation Research Board (TRB), and the Transit Development Corporation, Inc. (TDC), a nonprofit educational and research organization established by APTA. TDC is responsible for forming the independent governing board, designated as the TCRP Oversight and Project Selection (TOPS) Committee.

Research problem statements for TCRP are solicited periodically but may be submitted to TRB by anyone at any time. It is the responsibility of the TOPS Committee to formulate the research program by identifying the highest priority projects. As part of the evaluation, the TOPS Committee defines funding levels and expected products.

Once selected, each project is assigned to an expert panel, appointed by the Transportation Research Board. The panels prepare project statements (requests for proposals), select contractors, and provide technical guidance and counsel throughout the life of the project. The process for developing research problem statements and selecting research agencies has been used by TRB in managing cooperative research programs since 1962. As in other TRB activities, TCRP project panels serve voluntarily without compensation.

Because research cannot have the desired impact if products fail to reach the intended audience, special emphasis is placed on disseminating TCRP results to the intended endusers of the research: transit agencies, service providers, and suppliers. TRB provides a series of research reports, syntheses of transit practice, and other supporting material developed by TCRP research. APTA will arrange for workshops, training aids, field visits, and other activities to ensure that results are implemented by urban and rural transit industry practitioners.

The TCRP provides a forum where transit agencies can cooperatively address common operational problems. The TCRP results support and complement other ongoing transit research and training programs.

TCRP REPORT 11

Project C-5 FY '94
ISSN 1073-4872
ISBN 0-309-05714-0
Library of Congress Catalog Card No. 96-60355

Price $20.00

NOTICE

The project that is the subject of this report was a part of the Transit Cooperative Research Program conducted by the Transportation Research Board with the approval of the Governing Board of the National Research Council. Such approval reflects the Governing Board's judgment that the project concerned is appropriate with respect to both the purposes and resources of the National Research Council.

The members of the technical advisory panel selected to monitor this project and to review this report were chosen for recognized scholarly competence and with due consideration for the balance of disciplines appropriate to the project. The opinions and conclusions expressed or implied are those of the research agency that performed the research, and while they have been accepted as appropriate by the technical panel, they are not necessarily those of the Transportation Research Board, the Transit Development Corporation, the National Research Council, or the Federal Transit Administration of the U.S. Department of Transportation.

Each report is reviewed and accepted for publication by the technical panel according to procedures established and monitored by the Transportation Research Board Executive Committee and the Governing Board of the National Research Council.

Special Notice

The Transportation Research Board, the Transit Development Corporation, the National Research Council, and the Federal Transit Administration (sponsor of the Transit Cooperative Research Program) do not endorse products or manufacturers. Trade or manufacturers' names appear herein solely because they are considered essential to the clarity and completeness of the project reporting.

Published reports of the

TRANSIT COOPERATIVE RESEARCH PROGRAM

are available from:

Transportation Research Board
National Research Council
2101 Constitution Avenue, N.W.
Washington, D.C. 20418

Printed in the United States of America

FOREWORD

By Staff
Transportation Research Board

This report will be of interest to general managers, operations managers, and communications specialists responsible for communications systems within transit and paratransit organizations. The report provides information concerning the Federal Communications Commission (FCC) rules governing the refarming of the land mobile radio spectrum and their impacts on current and future transit communications systems requirements. The report contains a nontechnical executive overview (in Chapter 2) that describes the rules issued in June 1995 by the FCC regarding the refarming of radio frequencies, provides an overview of the impacts of radio frequency refarming, and offers potential courses of action for transit and paratransit systems. In addition, the report provides more detailed technical information for communications specialists, and includes several examples of potential cost impacts to transit and paratransit systems.

The Federal Communications Commission has decided to use "refarming" to help mitigate radio frequency congestion and increase spectrum efficiency in the private land mobile radio bands (frequencies below 512 MHz). Refarming is the term used for reduction in bandwidth allocated to radio channels in the designated bands. The refarming of frequencies has potential impacts on transit and paratransit communications systems and on capital procurement of communications equipment. FCC rules governing radio frequency refarming were issued in June 1995.

Under TCRP Project C-5, research was undertaken by Arthur D. Little, Inc. to assist transit and paratransit agencies in formulating plans for future actions in response to the FCC decision to refarm radio frequencies by providing information on the impacts of the FCC's rules.

To achieve the project objective, the researchers defined the scope of the FCC rules as they relate to the transit and paratransit industry. Once defined, the researchers characterized and assessed the impacts on the industry. To assist in this effort, a representative sample of transit and paratransit systems was surveyed to determine the types of communications systems and applications currently in use and planned for the future. Based on impacts identified for current and future communications-systems applications, a number of potential courses of action for transit and paratransit systems were identified for consideration in response to the FCC rules. Thus, the report is a valuable resource for transit managers and specialists responsible for communications systems within their organization.

CONTENTS

ACKNOWLEDGMENTS

Jeffrey J. Maul and John J. Greichen of Arthur D. Little, Inc., were the principal investigators for this project.

The TCRP Project C-5 panel, in particular Louis F. Sanders of the American Public Transit Association, provided important guidance and information enabling an efficient effort. The cooperation and insight of 21 transit survey participants were essential to a useful and accurate report. The principal investigators greatly appreciate the support of the project panel and survey participants.

CHAPTER 1

INTRODUCTION AND RESEARCH APPROACH

RESEARCH PROBLEM STATEMENT

The Federal Communications Commission (FCC) has decided to use "refarming" to help mitigate radio frequency (RF) congestion and increase spectrum efficiency in the private land mobile radio (PLMR) bands (frequencies below 512 MHz). Refarming will reduce the bandwidth allocated to radio channels in the designated bands. Base stations and mobile radio equipment operating within these bands may become obsolete if they cannot operate in the reduced bandwidth. The refarming of frequencies is likely to have a significant impact on transit communications systems and capital procurement of communications equipment in the near future. The FCC rules governing RF refarming were issued in June 1995.

OBJECTIVE

The objective of this research was to assist transit agencies in formulating plans for future actions in response to the FCC decision to refarm radio frequencies by providing information on the effects of the FCC's new rules. The research project defines the scope of the planned FCC changes as they relate to the transit industry and assesses the effects on the industry (e.g., communications capability and cost of compliance) by undertaking an inventory of current communications systems and functions.

RESEARCH PLAN

Task 1: Review FCC Plan

The FCC Notice of Proposed Rule Making (NPRM) related to "Refarming" (Docket No. 92-235) has four distinct objectives:

1. Reduce channel bandwidth (and spacing) for the land mobile ultra-high frequency (UHF) and very-high frequency (VHF) bands.
2. Implement "exclusive use overlay" to assign the new channels created.
3. Reduce transmitter power levels on the basis of Height Above Average Terrain (HAAT).
4. Set aside channels for "new technologies."

These objectives pose implications to the users of PLMR, particularly the transportation industry, which relies heavily on mobile communications. The FCC's effort is intended to reduce congestion and encourage more efficient use of the spectrum. This will enable more users to occupy the same band and not interfere with each other. On the other hand, this effort could cause service disruption and extreme financial effects because of forced obsolescence and premature retirement of equipment.

To perform Task 1, the research team visited the FCC's Private Radio Bureau (now known as the Wireless Telecommunications Bureau) to determine the latest developments, including the most recent schedule. The research team also investigated other industry perspectives by reviewing industry literature and contacting PLMR equipment suppliers. A degree of uncertainty remains on refarming, although initial rules have been issued. The goal during Task 1 was to understand the current rules and their resulting range of outcomes. The issuance of the refarming rules late in this project necessitated a second visit to the FCC to fully understand the Report and Order (R&O) issued on June 15, 1995.

The research team prepared a working paper describing the FCC Refarming NPRM, the status of the R&O, the schedule for implementation, and interpretive notations wherever appropriate.

Task 2: Survey Transit Systems

Task 2 of the research plan consisted of a survey of a representative sample of transit systems. The purpose of this survey was to determine the status of transit communication systems and applications and to understand plans for future systems and applications. The survey collected information that aided in assessing the effects of FCC refarming in Task 4.

The research team employed a proven approach to information collection in order to ensure that a complete and accurate survey was conducted. The following subtasks constituted the approach to this survey.

Identify and Screen Interviewees

The research team identified a balanced group of survey participants from the following categories:

- Type of transit system (e.g., bus, rail, or paratransit or taxicab),

- Location of system (e.g., urban, suburban, or rural), and
- Size of operation (e.g., large, medium, or small).

The proportions chosen for each of these categories were based on the understanding of the present and probable future use of mobile radio communications. This ensured that the transit operations most likely to be affected by refarming had a sufficient voice in the survey. the research team visited with personnel from the American Public Transit Association (APTA) and contacted the International Taxicab and Livery Association (ITLA) to receive guidance on adding smaller transit operations to the survey list. A preliminary listing of transit and paratransit and taxicab systems is shown in Table 1. The selected transit systems were then screened through brief phone interviews to ensure that mobile radio applications are employed or planned.

Create Detailed Survey Questionnaire

In parallel with the participant screening, the research team developed a detailed survey questionnaire (see Appendixes A and B). This questionnaire, which served as the primary guide for in-person and phone interviews, dealt with the present and future application of mobile radio technology and asked such questions as the following:

- What communications needs does the system have?
- What radio equipment is used? Manufacturers and models? How old is the average unit? How much have you invested in new equipment in the past 5 years?
- Describe your existing mobile radio system. (Trunking, simulcast, repeaters, mobile-to-mobile, mobile-to-dispatch, etc.)
- What are the usage scenarios in which communications are needed?
- What are the environmental conditions? (Background and RF noise, temperature)
- What are the frequencies in use? How many channels? How are they expected to change?
- What is the mix of analog and digital technology?
- What is the mix of voice and data communications required?
- What are the rate and duration of messages?
- What plans, if any, exist to deal with the FCC refarming issue? Are estimates of the cost impact available?
- What are the required characteristics of your mobile radio equipment? (Portability, battery life, number of channels, mobile repeater, talk-around, coded squelch, telephone interconnect, automatic number identification [ANI], etc.)
- What are the anticipated future requirements for the system? (Additional channels, future features, and future applications)
- Have you requested additional channels recently? Were you assigned the additional spectrum? If no, why not?
- Are you using other forms of wireless communications such as cellular telephone, specialized mobile radio

TABLE 1 Proposed survey participants

Transit System	Type and Size of Service					Location	Region	Notes
		Rail						
	Commuter	Heavy	Light	Bus	Paratransit			
Boston - MBTA	Medium	Medium	Large	Medium	Large	Urban,Suburban	East	Old System
New York - CTA		Large		Large		Urban,Suburban	East	
New York - LIRR	Large					Urban,Suburban	East	
Miami - Tri-Rail	Small					Suburban	East	
Washington - WMATA		Medium		Large		Urban, Suburban	East	New System
Los Angeles - SCRTD				Large		Urban,Suburban	West	Innovator
San Francisco - BART		Medium				Urban,Suburban	West	Innovator
Spokane, WA - TA				Small	Medium	Suburban,Rural	West	
Fresno,CA - TS				Small	Small	Rural	West	
Salt Lake City - TA				Medium	Small	Suburban	West	
St. Louis - Bi-State			Small	Medium	Small	Suburban, Rural	Central	New System
Austin,TX - Capital MTA				Small	Medium	Suburban	Central	
New Orleans - RTA			Medium		Small	Urban, Suburban	Central	
Cleveland - RTA		Small		Medium	Medium	Urban,Suburban	Central	

(SMR), or citizen's band (CB)? Do you plan to? Are they in lieu of PLMR? If yes, why?

Several of these questions assume that transit system personnel have begun to consider options relative to refarming. The research team believes it is useful to determine what action plans may be emerging, potentially supplementing action plans identified as part of this research.

Conduct Survey

The research team distributed the survey to all prescreened participants and scheduled an agreeable time for the detailed discussion. Most discussions took place by telephone, with local or regional discussions in person when possible. The results of the survey were synthesized into a summary format and reviewed for inconsistencies. Follow-up phone discussions were undertaken to clarify inconsistencies. The output of this survey task was a working paper that presented a profile of the transit industry's current and planned communications systems and applications.

Task 3: Compile Technology Listing

The research team contacted the dominant suppliers in the industry and determined what products they intend to introduce in response to FCC refarming. The research team also performed an industrywide search of literature and research and contacted industry pundits in order to identify recent technological developments that may affect mobile radio systems and mobile radio equipment as they relate to refarming. In addition, the research team investigated organizational attempts to influence or establish standards, such as the Association of Public-Safety Communications Officials (APCO) Project 25 or the National Telecommunication and Information Administration (NTIA) narrowband standards, to determine how these attempts may affect refarming and the available equipment alternatives.

After gathering the technology-related information, the research team compiled a list of the specific technologies and developments, both established and impending, that relate to or may affect refarming. The research team itemized the advantages and disadvantages of each and identified any technological conflicts.

Task 4: Determine Impacts

FCC Docket No. 92-235 has been controversial. It may require an extremely large population of radio users, many of whom are responsible for maintaining public safety and protecting property, to adopt emerging technology.

The primary objective of refarming is to make the spectrum available to new users and applications, but this may have financial and equipment effects for those already heavily relying on radio service. On the positive side, refarming promises to boost mobile radio service to a new level of technology that may enable users to solve longstanding channel contention and interference problems, take advantage of new spectrum to add additional channels, and benefit by new features inherent in the latest digital equipment.

The research team also recognize that transit systems requirements for wireless communications are increasing and will continue to increase with the advent of new applications. The analysis took into account present transit communications requirements and those related to impending transit requirements, such as those identified in the National Program Plan for Intelligent Vehicle Highway System (IVHS) (now called the Intelligent Transportation System [ITS]), including the following:

- Automatic vehicle location (AVL);
- Vehicle and driver performance monitoring;
- Security and emergency request systems;
- Information systems that convey route and schedule changes, navigational and electronic maps, and rider information;
- Collision warning and avoidance systems;
- Electronic payment services for tolls, fares, and parking; and
- Travel information centers (kiosks).

The risks involved are significant, and the potential effects are numerous. The following is a partial list of potential effects that must be considered:

- Early retirement of working, reliable, known equipment;
- Selection of new equipment (and technology) from a relatively small number of available products;
- Commitment to a single vendor because the equipment procured probably will be proprietary;
- Implementation of an unknown technology;
- Coexistence with other mobile radio operators who also are implementing "unknown" technology and can interfere with one's system;
- Maintenance of a new system, including service training, unfamiliar documentation, new test equipment, spare parts, and new vendor relationships;
- User or operator training issues (i.e., new operational procedures, feature orientation, additional channels, and modified coverage);
- User acceptance of the new system, in light of delays in transmission or distorted speech caused by low bit-rate compression schemes;
- Smaller geographical coverage because of new power limitations on the system, which may necessitate the addition of base stations, and the accompanying logistics; and
- Operational conflicts caused by equipment from differ-

ent suppliers operating on the same or closely spaced channels.

The research team prepared a report identifying the effect that refarming will or may cause and, when feasible, quantified and qualified the effect. The research team also reported the relevant, known consequences associated with each effect.

Task 5: Describe Possible Actions

Task 5 of the research plan consisted of assessing and describing strategies that transit agencies may implement to adopt frequency refarming as proposed by the FCC. These strategies may include recommendations for further research, if appropriate.

The approach to this task was to have the research team "brainstorm" possible actions or research. These brainstorming sessions benefited from the information gathered in Tasks 1 through 4. Concepts were developed that addressed the various situations that were derived from the survey and assessing the cost and operational issues that might exist. Methods for migrating to the FCC compliance over varying time frames were explored. Further research beyond this project is warranted, given the 10-year implementation required by the FCC.

At the completion of this task, the research team submitted to TRB a working paper describing the definitive action plans and recommending continuing research.

Task 6: Recommend Dissemination Plan

Task 6 consisted of developing a plan to disseminate the results of this research project to the transit industry. The research team assessed how various media and forums could be used to maximize the exposure for this research. An early element of this dissemination was a presentation of results at the 1995 APTA annual meeting. The research team assessed the value of placing articles in periodicals such as *Mass Transit*, *Railway Age*, *Radio Resource*, *Mobile Radio Technology*, *APCO Bulletin*, *Communications*, or *Radio Communications Report*. Presentations at conferences sponsored by organizations such as the following should be considered:

- U.S. Department of Transportation (DOT) and its agencies,
- ITS program,
- Advanced Public Transportation Systems (APTS) program,
- Advanced Train Control Systems program,
- Association of American Railroads (AAR),
- Busing associations,
- State and regional transit agencies,
- *Railway Age's* Communications and Signalling Conference, and
- ITLA.

The result of this task was a working paper that described the audience for the research, recommended publications and forums for dissemination, and provided a time frame for implementation.

Task 7: Prepare Final Report

In Task 7, the research team produced a final report documenting the approach to and outcome of the research. The working papers developed during each task of this effort served as the basis for the final report. Information developed during subsequent phases was incorporated in the final version of each paper as appropriate.

CHAPTER 2

FINDINGS

EXECUTIVE OVERVIEW

This executive overview discusses key findings of TRB TCRP Project C-5. This overview is nontechnical and does not include all effects and action alternatives defined in the detailed report.

Radio System Usage in Transit

Transit systems in the United States today depend on radio-based communication systems to help deliver efficient, safe, and reliable service to millions of riders. Radio systems are an essential tool for rail engineers, drivers, maintenance, security, and management personnel. Such systems enable instantaneous access to people and exchange of information within the service area.

Radio systems traditionally are used for person-to-person voice communications only; however, transit operators are becoming interested in emerging data applications involving person-to-computer or computer-to-computer communications, such as real-time monitoring and control of equipment and access to information databases. As a result, the already heavy reliance on radio communications will increase.

Many transit systems are experiencing overcrowding or congestion in their radio systems, which is affecting their operations adversely. Overcrowding is a particular issue for urban transit systems, where the radio spectrum is sought after by many businesses. The future data-intensive uses for radio systems will only worsen this situation. The classic approach to relieving congestion is to request additional radio channels from the FCC. Today, however, in many areas additional channels are not available because of overcrowding.

FCC's Solution to Overcrowding

The FCC has been working for the past few years on methods to alleviate spectrum overcrowding. The Commission has learned that, because of the diverse and divergent needs of the radio system users, there is no simple solution. The Commission has also recognized that the radio spectrum is a valuable commodity that must be allocated carefully. Recent personal communications services (PCS) spectrum auctions held by the FCC have realized a windfall of more than $14 billion.

In June 1995, the FCC chose to implement a plan to refarm the PLMR spectrum, replacing wideband radio channels with narrowband channels to increase the number of users who may occupy the spectrum simultaneously. To accomplish this, the users must rely on new technology to maintain adequate performance in these narrowband channels.

The transition from the current wideband channels of 25 KHz to narrower-band channels will occur in two stages as follows:

- Stage 1—August 1, 1996 (12.5 KHz) and
- Stage 2—January 1, 2005 (6.25 KHz).

Contrary to earlier proposals made by the FCC, these new rules apply to equipment manufacturers and their new products—not directly to mobile radio system users such as transit operators. Wideband-channel radio products will continue to be available as long as suppliers wish to offer them, and users will not be prohibited from using these products. Transition to narrowband channels will occur at the pace defined by the user.

The FCC has not completed its plan for the reallocation of the radio spectrum. The FCC is likely to impose fees or auctions on transit operators. Even higher economic penalties will be incurred by those transit operators who do not adopt narrowband-channel technology. The transit industry could be faced with decreased channel access, even if fees are not imposed. New channels created through refarming may be sold to the highest bidder. Discussion between the industry and the FCC has just begun on this matter, and it will be at least a year before specific rules are released.

Radio Technology for Narrower-Band Channels

Radio equipment manufacturers probably will continue to offer wideband-channel products well into the future. A broad market for these products exists in transit, public safety, and many other applications.

Narrowband-channel equipment compatible with Stage 1 requirements (12.5 KHz) is available from several manufacturers. Stage 1 equipment is 10 to 40 percent more expensive than wideband-channel equipment, especially digital products (as opposed to narrowband-channel analog products). The digital radios offered today also do not meet certain performance criteria of their analog counterparts. These considerations concern users and are partly responsible for the current limited deployment of digital products.

Stage 2 narrowband-channel equipment is not available and has yet to be demonstrated successfully in the channels used by transit operators. Acceptable performance in these narrower-band channels will require a technological breakthrough. Equipment manufacturers now believe a breakthrough can be accomplished in time to meet the deadline in the year 2005. The costs of these future products are not known.

Effect of FCC Rules on Transit

On a positive note, the FCC rules eventually should relieve the overcrowding of transit radio channels, assuming transit is allowed to retain the necessary frequency allocations. The rules will create up to four times the number of channels currently available.

The disadvantages of these new rules may outweigh the advantages. The new FCC rules undoubtedly will result in higher costs for the transit industry. The questions are how much and when. The answers to these questions depend on the answers to the following questions:

- **Can transit stay just with wideband channels?** Wideband-channel equipment may become less available or increase in cost. Fees imposed for the right to use channels that previously were free may greatly increase the differential cost of continuing to use wideband channels.
- **What will the wideband- and narrowband-channel radio equipment pricing and availability be in the future?** Narrowband-channel equipment will cost 10 to 40 percent more than current equipment for several years after its introduction. Vendors may discontinue older wideband-channel equipment or raise the cost considerably as the market diminishes.
- **Will narrowband-channel technology have performance equal to current systems?** Performance probably will be lower and additional equipment (e.g., additional base sites to maintain existing coverage) may be needed to compensate.
- **What are the installation, training, and maintenance effects of the transition to Stage 1 and Stage 2?** Costs will increase, at least temporarily, during each transition to get the system running smoothly again.
- **Will there be more than one vendor to buy from?** As major equipment providers pursue different approaches to narrowband-channel equipment, procurement options probably will be limited. Other vendors will require time to catch up to the major providers, thereby limiting the competitive bidder list. Suppliers may not feel the need to be competitive for spare parts, service, and system add-ons, once the user has committed to their proprietary technology.

Although the research team does not have more specific answers to these questions today, the team is certain the answers will mean higher acquisition and operational costs for narrowband-channel radio systems or higher costs for wideband-channel systems through higher spectrum fees, maintenance, or system expansion costs. Transit systems in need of new radio systems are particularly at risk of increased costs. A newly purchased wideband-channel system may have to be replaced before the end of its useful life. Even a Stage 1 narrowband-channel system may have to be largely replaced within 10 years as the Stage 2 fees are enforced.

Actions for Transit

The transit industry should act to protect its interests and to influence proposed FCC rules. Unfortunately, considering the options available, few actions will avoid higher costs. Transit systems can only hope to minimize the total cost effect of the new rules. Possible options for transit operators in response to the rules the FCC has issued include the following:

- **Defer radio system investments until the situation settles.** The rules might change somewhat if enough opposition is heard, but major changes should not be expected. Equipment vendors may devise more desirable solutions, given time to further develop narrowband-channel technology.
- **Buy radio systems that are not subject to the new rules.** Certain higher frequency radio systems are not required (as yet) to move to narrowband channels. Transit operators could buy one of these systems and avoid the new FCC rules. The caveat is that these higher frequency systems are up to 50 percent more expensive than current systems.
- **Rent a radio system instead of purchasing one.** Transit operators can rent equipment and channels from another operator, public or private. This action can enable a transit operator to augment and delay retirement of a radio system that no longer can meet the needs of the transit operation.
- **Proactively influence the proposed fee or auction proposal.** Discussion has just begun on the concept of fees or auctions. The transit industry should have its voice and interests heard through organizations such as APTA and ITLA.

FCC RF SPECTRUM REFARMING

Background

PLMR supports more than 12 million radio users, including most public transit organizations. PLMR users depend heavily on radio communications to perform their work or business, and the country as a whole relies on radio communications to protect lives and property.

For the last 70 years, PLMR has been one of the largest, most important services regulated by the FCC; however, the existing regulatory rules are somewhat dated and based on old technology and regulatory concepts. Many PLMR channels have been unacceptably crowded for years, and many geographic population centers have experienced or soon will experience significant spectrum shortages.

The FCC has performed some studies, each confirming that the available spectrum is either overcrowded or soon will be exhausted. This condition has caused some legitimate PLMR users to seek alternative communications mechanisms and has hindered the ability of many users to achieve the full benefit of their systems. The FCC has instituted several rule changes recently, such as adding spectrum and combining services, in order to ease the situation. Although these changes have all had positive effects, the cumulative effect falls well short of the need.

In response, the FCC Wireless Telecommunications Bureau has proposed a set of regulatory changes with the objective of updating the rules to support future technologies, specifically technologies tailored to provide relief for the spectrum congestion. (Refer to Appendix C for definitions of PLMR and FCC terminology.)

FCC NPRM

In October 1992, the FCC issued an NPRM, Docket No. 92-235, containing a comprehensive set of proposals reflecting comments received from a Notice of Inquiry (NOI) issued earlier. The FCC was convinced that without significant regulatory changes in the bands below 512 MHz, the quality of PLMR communications probably would deteriorate to the point of endangering public safety and the national economy.

The stated objectives of the NPRM were as follows:

- Increase channel capacity in the bands allocated to PLMR,
- Promote more efficient use of these bands through frequency reuse and other mechanisms, and
- Simplify the regulatory rules and policies governing the use of these bands.

The resulting NPRM was comprehensive and detailed. The major proposals within the NPRM were as follows:

- **Establish spectrum efficiency standards** requiring that radios operate on narrower-band channels, initially in 12.5-KHz channels, progressing to 6.25 KHz;
- **Allow exclusive use of channels** by licensees meeting the usage loading criteria, in the bands above 150 MHz, to encourage the adoption of more efficient technology;
- **Consolidate the current 19 radio services** to ease frequency management and improve channel use;
- **Limit transmitter power and antenna height** to restrict excessive radio transmitter emissions, and thereby increase frequency reuse; and
- **Create an innovative shared use service category** by allocating a band of channels.

These proposals and others presented in the NPRM were included in a simplified set of regulatory rules, Part 88, which were proposed to replace the existing Part 90 of the Code of Federal Regulations (CFR). These proposals are described in detail in the following sections.

Establish Spectrum Efficiency Standards

There are three ways to create more radio channels: add more spectrum, reduce the amount of spectrum required for each channel, or find a way to reuse the available spectrum. Because no additional spectrum is available, only the latter two options are viable, and the FCC proposed to implement both. Reducing the amount of spectrum per channel was addressed by new spectrum efficiency standards, and frequency reuse was addressed by limiting the transmitter power of stations.

To provide the technical flexibility to permit the use of emerging technologies, the FCC chose to specify spectrum efficiency relative to narrowband (i.e., radios that use a narrower-band channel or a fraction of an existing channel) as a benchmark. Thus, although the spectrum efficiency requirement could be met by deploying narrowband radio technology, other technologies could be deployed as long as they achieved at least the same efficiency. This would allow the economic and public safety considerations to determine the best technology for each application while requiring that PLMR allocations be used efficiently.

The standards proposed provided for greater efficiency over time and varied depending on the frequency band of operation. The schedule of implementation would occur in two stages. The first stage would require users to reduce the occupied bandwidth of their equipment by a specified date. The FCC stated that existing equipment could be easily retuned to operate within 12.5-KHz bandwidth for minimal cost. This process was expected to create 2,200 to 3,100 new narrowband channels between the existing channels. The second stage would require the users to deploy new spectrum-efficient equipment, further reducing the occupied bandwidth.

To encourage the rapid adoption of new spectrum-efficient equipment, the FCC offered licensees the opportunity to retain two narrowband channels for every channel, by implementing spectrum-efficient technology at least 2 years sooner than required by the rules. Furthermore, the FCC stated that licensees could fund the equipment conversion by reassigning part of this spectrum to third parties willing to reimburse them. The proposed schedule of implementation of the new efficiency standards is shown in Table 2.

Allow Exclusive Use of Channels

PLMR rules for bands below 470 MHz would not provide for exclusivity. That is, users would be assigned a channel on a shared use basis. The number of users on a channel would be based on the loading offered by each user and would vary by market and service.

Licensees operating in a shared use channel would have little or no incentive to adopt spectrum-efficient technology. Furthermore, all users of a channel would have to agree to adopt spectrum-efficient technology to achieve compatibility.

In response, the FCC has proposed the use of a marketplace mechanism called exclusive use overlay (EUO), which would enable licensees with sufficient channel loading to protect their radio environment by converting shared use channels to exclusive use channels. This proposal would provide a temporary freeze on new licensing in specific channels in specific locations, if applicants obtained concurrence from large licensees. Once concurrence of all large licensees was achieved, the FCC would permanently freeze licensing. No additional use of the particular channel within 50 miles would be permitted without concurrence of the EUO licensee. To receive an EUO license for a channel without current licensees, the applicant would have to meet the loading requirement within 8 months of its EUO authorization.

It was believed that the EUO would provide the licensees with the incentive and the opportunity to protect and conserve the spectrum. The FCC would leave a significant number of channels available for licensing on the traditional shared use basis.

Consolidate the Current 19 Radio Services

Part 90 of the CFR divides the PLMR channels into services, which group licensees by the service that the user performs. Examples of services are Police Radio Service, Fire Radio Service, Forestry Radio Service, and Railroad Radio Service. There are 19 PLMR radio services, meaning that the available PLMR radio channels are divided into 19 groups. Licensees are only eligible for channels assigned to

TABLE 2 1992 FCC proposed Part 88 transition plan

	Effective Date of Requirement			
	Effective date of new rules (assumed to be Jan 1994)	Jan. 1996	Jan. 2004	Jan. 2005-2012
72-76 MHz	All new systems must use 5-KHz channels and conform to power and HAAT limits. Existing licensees can voluntarily adopt 15-KHz channelization.	Reduce deviation to 15-Hz occupied bandwidth and conform to power and HAAT limits.	Top 15 markets must convert to 5-Hz channels.	Smaller markets must convert to 5 MHz on graduated schedule.
150-174 MHz	All new systems must use 5-KHz channels and conform to power and HAAT limits. Existing licensees can voluntarily adopt 15-KHz channelization.	Reduce deviation to 15-KHz occupied bandwidth and conform to power and HAAT limits.	Top 15 markets must convert to 5-MHz channels.	Smaller markets must convert to 5 MHz on graduated schedule.
421-512 MHz	All new licenses must use 6.25-KHz channels and conform to power and HAAT limits. Existing licensees can voluntarily adopt 12.5-KHz channelization.	Reduce deviation to 12.5-KHz occupied bandwidth and conform to power and HAAT limits.	Top 15 markets must convert to 6.25-MHz channels.	Smaller markets must convert to 6.25 MHz on graduated schedule.

their radio service or they must apply for intercategory sharing, which has been criticized as being ineffective.

As a consequence, channel use is not consistent across the 19 user groups. A 1992 study showed variations in use often exceeding factors of 10 for channels in the same frequency band designated for use by different radio services. The study, however, did not measure frequency use, so it is not clear if there is any disparity between the radio services.

The FCC believes that some consolidation of the current alignment of radio services is necessary to realize maximum use of the PLMR spectrum. The Commission proposed the following alternatives:

1. Consolidate the radio services into three broad categories: Public Safety, Non-Commercial (i.e., entities that use the radio service for internal use), and Specialized Mobile Radio (i.e., commercial), and a General Category Pool encompassing all three services.
2. Retain the current 19 services but assign all new frequencies to the proposed new broad categories as defined in Alternative 1.

The FCC stated it did not have a preference for either option, and it invited comments and alternative proposals.

Limit Transmitter Power and Antenna Height

Geographic frequency reuse is a very effective way to improve spectrum use. This concept has received considerable attention recently because it is the basis of cellular telephone service. Frequency reuse implies limiting the power of the transmitted signal to confine the signal to the desired geographical area and to constrain the signal spillover in neighboring geographical areas where the licensee does not need coverage. The result is that the same radio channel can be reused by a different licensee in the neighboring geographical area.

The range or distance reached by a radio transmitter is a function of the transmitter output power, the frequency of the transmitter, the antenna gain, and the antenna height. Effective radiated power (ERP) takes into account transmitter power and antenna gain. Believing that many current licensees use far more power than required, the FCC feels compelled to adopt restrictions on transmitted power levels. The FCC proposed a maximum authorized ERP of 300 watts, in the 150- to 174- and the 450- to 470-MHz bands, for stations with an antenna HAAT of up to 60 m (197 ft) and lower ERP levels for stations with antenna HAAT of more than 60 m. The FCC believed this would enable reuse of a frequency at a distance of approximately 80 km (50 mi) from the licensed station.

All systems in the 150- to 174- and 450- to 470-MHz bands were to have met these more stringent power and antenna height and bandwidth limitations by January 1, 1996. In addition, after the effective date of the new Part 88 (which was presumed to be January 1, 1996), any trunked channel, new channel, new site, or system with an EUO license older than 6 months would have to meet the new standards.

Create an Innovative Shared Use Service Category

The FCC proposed to designate approximately 250 channel pairs in the 150- to 162-MHz band for a new, wide-area, spectrum-efficient, voice and data communications service. The Commission proposed granting five licenses via lottery in each of seven regional markets for this new type of shared use radio operations.

Industry Comments on the Proposed Rules

The response from industry, both users and vendors, generally supported the refarming concept but strongly opposed the FCC's specific approach. (Copies of specific industry comments and the NPRM itself can be obtained by contacting the FCC Wireless Telecommunications Bureau at 202-634-2443. The Bureau can identify local copy services through which copies can be obtained.) Most respondents considered the schedule of implementation, the cost of compliance, and the technical uncertainty as presenting unacceptable risks. On the other hand, there was no consensus on an alternative proposal. The Land Mobile Communications Council (LMCC) responded with a Consensus Plan that received the most attention from the users and the FCC. This Consensus Plan and other key industry responses are summarized below.

Comments of LMCC

LMCC is a nonprofit association representing users of land mobile radio (PLMR), providers of land mobile services, and manufacturers of PLMR equipment. LMCC submitted an alternative Consensus Plan in response to the NPRM. The salient issues addressed in the Consensus Plan are summarized here.

Establish Spectrum Efficiency Standards. LMCC essentially agreed with the FCC's proposal to introduce spectrum efficiency standards. LMCC did, however, propose an alternate implementation schedule and process. For example, LMCC proposed the following schedule for the 421- to 512-MHz bands:

- January 1, 1994 (assumed to be the effective date of the new rules), existing licensees on full-power channels

would have the option of employing true 12.5-KHz bandwidth on a voluntary basis.
- January 1, 1994, a 6.25-KHz channel plan would be incorporated into the rules for voluntary use by licensees.
- January 1, 1994, a percentage of the current offset channels would be designated as primary channels.
- January 1, 1996, all equipment type-accepted (approved for use by the FCC) must operate on 12.5-KHz channels.
- January 1, 1999, the FCC would reconsider whether to require conversion to 6.25-KHz channels by January 1, 2014.
- January 1, 2004, all systems not employing 12.5-KHz equipment would be permitted to operate only on a noninterfering basis.

LMCC proposed two options for channelization of the 150- to 174-MHz band. Option A is similar to the 421- to 512-MHz plan outlined previously. Option B forgoes the interim conversion to 12.5 KHz and focuses directly on a process for introducing 6.25-KHz channelization. The common concerns addressed by each option were the short migration time and unproven narrowband technology.

Offer Exclusive Use of Channels. LMCC supported the introduction of exclusive channel assignments in the 150- to 174-MHz and 421- to 512-MHz bands.

Limit Output Power and Antenna Height. LMCC agreed that licensees should be required to limit signal power but proposed that system power levels should be commensurate with licensee service area requirements as follows:

- A "safe harbor" table of permissible ERP-HAAT combinations was proposed.
- Alternately, power should be limited to the minimum required to cover the authorized service area on the basis of calculated range.

Create an Innovative Shared Use Service Category. LMCC urged the FCC not to adopt the Innovative Shared Use proposal. It was argued that this proposal was not in the public interest and that the channels in question could be used more effectively for traditional and advanced technology land mobile systems.

Comments of AAR

AAR is the primary railroad association and serves as frequency coordinator for the Railroad Radio Service. AAR participated in creating the LMCC Consensus Plan and offered the following additional comments:

- Retain the Railroad Radio Service and grant exclusivity for all channels currently in the Railroad Radio Service.
- Retain AAR as sole coordinator for the Railroad Radio Service.
- Eliminate the requirement to retune transmitters for reduced deviation.
- Define an offset-overlay for the VHF band used by railroads, on the basis of 12.5-KHz bandwidth and 7.5-KHz channel separation.
- Allow an alternate migration plan and a unique timetable controlled by the railroads.

Comments of APCO

APCO represents the interests of the public safety PLMR community and is the certified frequency coordinator for police, local government, and public safety.

APCO participated in developing LMCC's Consensus Plan and, therefore, agreed with the basic recommendations outlined within the plan. APCO did, however, express opinions unique to its interests, including the following:

- The FCC should establish separate regulations for public safety in order to consider special needs.
- Public safety has a severe spectrum shortage and requires more channels than refarming will provide.
- APCO believes that all users of the spectrum must refarm, and it is particularly concerned with the television broadcasters' use of spectrum.
- Any newly created channels should remain in the same radio service.
- Because public safety communications are critical, it is not wise to intersperse public safety channels with other communications signals that could interfere with or degrade the performance of public safety communications.
- APCO urges the FCC to abandon its proposed HAAT-based power and height limitations. Instead, APCO proposes that the FCC permit the frequency coordinator (APCO) to limit coverage to the user's jurisdictional boundary.
- APCO is concerned that the FCC will affect its Project 25 standard and its goal of creating an interoperable digital standard.

Comments of a Sampling of Equipment Vendors

The equipment vendor community generally supported the LMCC Consensus Plan but voiced serious concerns and offered recommendations about specific aspects of the NPRM, including the following:

- Very narrow narrowband technology (6.25-KHz and 5-KHz channelization) has yet to be proven in a real-world environment.
- The trend is toward new wireless applications in which more, not less, bandwidth is required to access greater amounts of information.
- The migration plan would cause significant destructive and harmful interference to users. The FCC's assumption that radio transmitters may simply be retuned as a first step in the spectrum efficiency plan is technically not possible. It has been stated that the performance of these radios would be seriously degraded by this retuning, and any attempt to use the vacated spectrum for other transmissions would cause serious interference to users.
- A longer amortization and transition period is needed for licensees. Users in rural areas where little spectrum congestion exists could continue using current equipment on a noninterference basis past the amortization period.
- Any decision to require conversion to band channels narrower than 12.5 KHz should be deferred, pending field experience in implementing narrowband equipment at 220 MHz.
- The FCC requirements would unduly limit a user's choice, because few narrowband products would be available.

FCC Decision and Status

The research team assessed the FCC decision and status of the refarming proceedings through discussions with the FCC (two visits), user associations, and equipment vendors. Since the 1992 NPRM, the significant concerns raised by the user community have resulted in a reassessment by the FCC. The spacing of frequency channels and timing for the transition remained the primary impediments to progress.

Although widely endorsed, the LMCC Consensus Plan did not prove completely satisfactory to the FCC or certain user groups. A new "Consensus User Group," which included AAR, APCO, the American Petroleum Institute (API), AASHTO, the American Trucking Association (ATA), the Industrial Telecommunications Association (ITA), the Manufacturers' Radio Frequency Advisory Committee (MRFAC), the National Association of Business and Educational Radio (NABER), and the Utilities Telecommunications Council (UTC) drafted an alternate proposal. This proposal also proved unsatisfactory to the FCC.

The FCC released an R&O and a Further Notice of Proposed Rule Making (FNPRM) on refarming on June 23, 1995. The R&O implemented some of the proposed changes and revised, eliminated, or postponed action on others. The R&O, according to the FCC, meets the intended goal of promoting spectrum efficiency while minimizing the adverse effect on users.

The FNPRM opens up a new docket dealing with the most difficult issues, such as exclusivity and channel assignment. The FNPRM also raised the subject of fees or auctions for PLMR channels. The FCC's success in auctioning frequencies in other bands has caused it to consider auctions as an effective impetus for spectrum efficiency. The FCC now considers market-based user fees and competitive bidding to be more effective than mandating spectrum efficiency standards. This report does not deal with the FNPRM and considers only the effects of the R&O.

The major rulings detailed in the R&O are the following:

- Channelization — New channels will be established while retaining the current channel centers to allow users to remain on their licensed frequency throughout a transition to narrowband. Retaining "on-channel" was seen as critical to licensees. The current 15-KHz channels in the 150- to 174-MHz VHF band and 25-KHz channels in the 421- to 430-, 450- to 470-, and 470- to 512-UHF bands will become two 7.5-KHz VHF and four 6.25-KHz UHF channels, respectively. Figure 1 illustrates the adopted UHF band plan and Figure 2 illustrates channel migration options for UHF. The channel plan is technology neutral, allowing narrowband-equivalent equipment (such as time division multiple access [TDMA]) to aggregate narrowband channels and still meet requirements.
- Transition to Narrowband—The FCC will manage the transition to more efficient use of the PLMR spectrum through the type-acceptance process. Future equipment must meet increasingly efficient standards over a 10-year period. The transition dates and conditions for the type-acceptance rules are as follows:
    - August 1, 1996 — New type-accepted equipment must be designed to operate on channels of 12.5 KHz or less or on channels of 25 KHz (if the narrowband efficiency standard is met). Multimode equipment that operates on 25-KHz channels will be allowed if it can operate on 12.5-KHz or narrower-band channels.
    - January 1, 2005 — New type-accepted equipment must be designed to operate on channels of 6.25 KHz or less or on channels of up to 25 KHz (if the narrowband efficiency standard is met). Multimode equipment that operates on 25-KHz or 12.5-KHz channels or both will be allowed if it can operate on 6.25-KHz or narrower-band channels.

Thus, equipment vendors must respond to the rules for new designs, while users have the flexibility to make the transition to narrowband at their own pace. Wideband equipment that is type-accepted before the

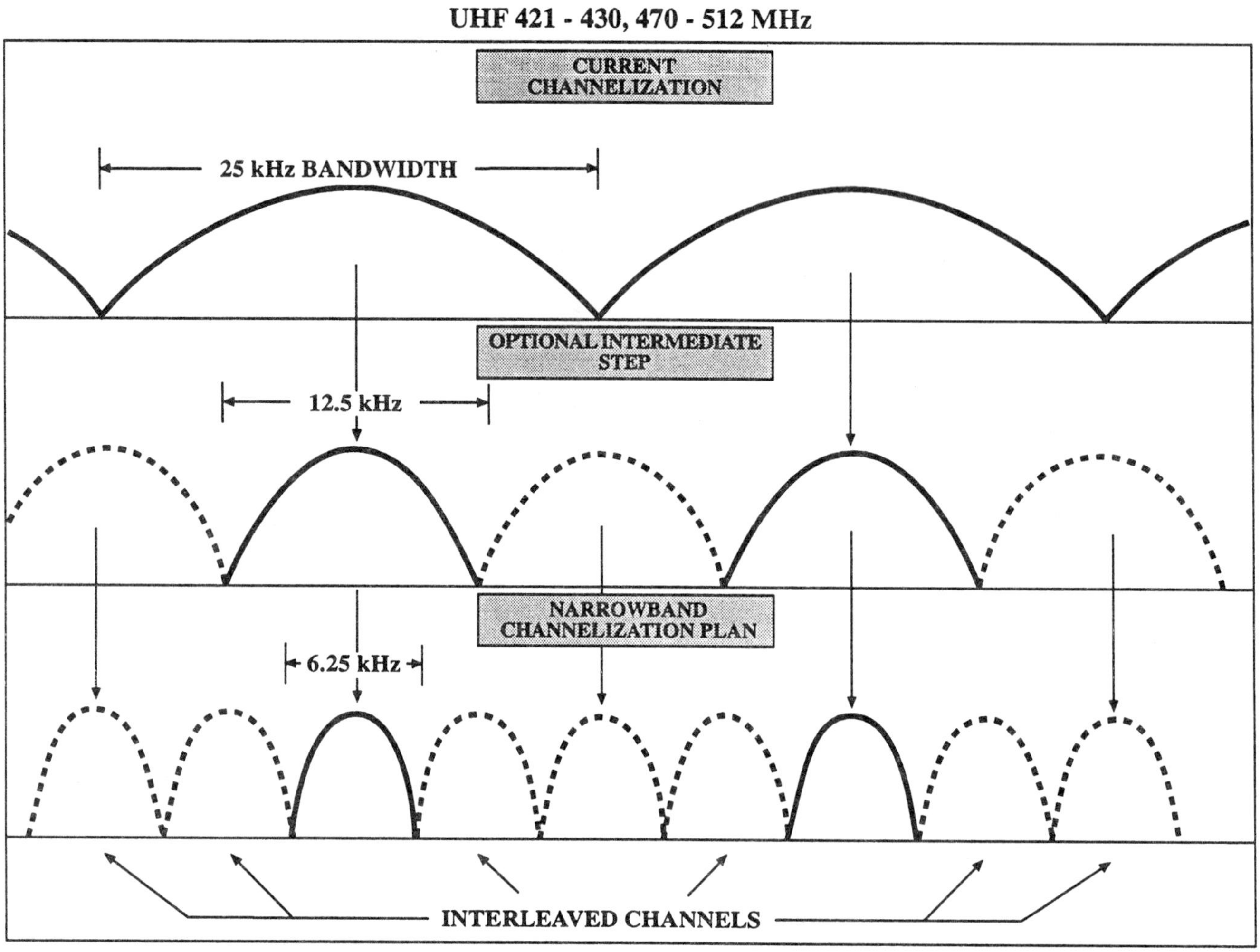

Figure 1. Adopted UHF Band Plan.

transition dates may continue to be manufactured and used indefinitely. The FCC believes that PLMR users will choose to replace retiring equipment with dual-mode equipment available from major vendors. Dual-mode equipment can function at wideband and narrowband. For example, Motorola sells a product that can function at either 25 KHz or 12.5 KHz. A user can eventually replace all wideband equipment with dual-mode equipment and then convert the entire system to narrowband. The FCC assumes the price differential between single-mode and dual-mode products will be minimal. An additional impetus for purchasing dual-mode products will be the awareness of future economic incentives (e.g., fees or auctions) for spectrum efficiency resulting from the FNPRM.

- Consolidation of Radio Services—The FCC feels consolidation is desirable to ensure more efficient distribution of the additional channels created by the transition to narrowband. The Commission's goal is to distribute spectrum use efficiently across the PLMR marketplace by creating competition in the frequency coordinator function. The FCC offered the example of enabling use of forestry frequencies for nonforestry purposes in Manhattan. The Forestry Radio Service controls and unnecessarily withholds those frequencies, according to the FCC. The FCC has asked the PLMR community to negotiate and propose a plan within 3 months of the date of the R&O. The FCC will then issue final rules on consolidation approximately 3 months later. A consensus on consolidation does not exist within the FCC, but this situation should not delay a ruling.
- Safe Harbor Power/HAAT Table—The FCC adopted a modified version of the LMCC safe harbor power limitation approach. The FCC limited transmitter output power for frequencies below 76 MHz is as follows:

- <25 MHz - 1,000 watts
- 25-50 MHz - 300 watts
- 72-76 MHz - 300 watts

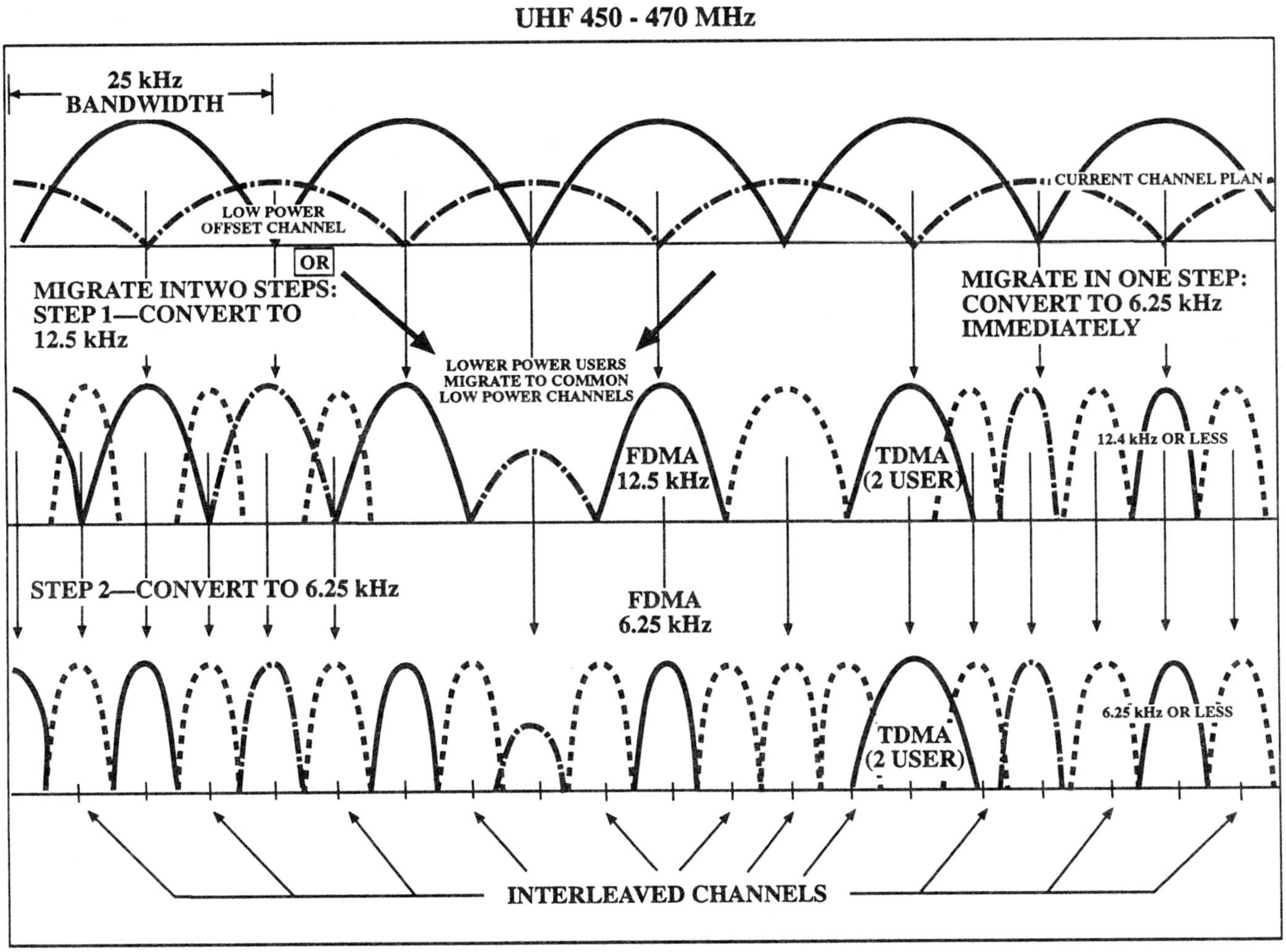

Figure 2. Channel Migration Options.

Limitations for the 150- to 174- and 450- to 470-MHz bands are defined in Tables 3 and 4. The limitations for the remaining bands (i.e., 220 to 222, 421 to 430, and 470 to 512 MHz) were not revised by the R&O. Users may apply for higher ERPs than shown in the tables, provided an analysis is performed to verify that signal levels are below a threshold at the borders of the service area. These new limitations apply only to new stations, a major deviation from the original proposed rules. However, any system transition to narrowband, relocation of base stations, and certain other activities is considered a "new" system by the FCC and becomes subject to safe harbor rules. The research team believes this interpretation of new systems may be protested and will become a subject for reconsideration.

- Other Technical Parameters—The FCC also specified rules on certain other technical parameters, including channel bandwidths, emission masks, and frequency stability.

The Refarming R&O did not implement the following:

- Innovative Shared Use Service Opportunities—Public safety organizations offered the most vocal opposition on this issue. They were concerned that mixing non-public safety with public safety channels offered the potential for interference between channels during critical operations, endangering the operation and the public. Public transit organizations can claim a similar operational concern.
- Deviation Revision—Users and manufacturers have effectively proven to the FCC that a deviation "adjustment" is not practical.

A major uncertainty surrounding the R&O is the issuance of new frequencies in the near term. The FCC indicated it was not certain how consolidation and the creation of new narrowband channels would affect new applications for frequencies. The FCC indicates generally favorable feedback to the R&O as of late August 1995 and does not expect significant reconsideration.

TABLE 3 150–174-MHz ERP/HAAT limitations

150 to 174 MHz - Maximum ERP/Reference HAAT for a Specific Service Area Radius										
Service area radius (km)	3	8	13	16	24	32	40	48[4]	64[4]	80[4]
Maximum ERP (w)[1]	1	28	178	500[2]	500[2]	500[2]	500[2]	500[2]	500[2]	500[2]
Up to reference HAAT (m)[3]	15	15	15	15	33	65	110	160	380	670

[1] Maximum ERP indicated provides for a 37-dBu signal strength at the edge of the service area per FCC Report R-6602, Fig. 19 (see § 73.699, Fig. 10).

[2] Maximum ERP of 500 watts is allowed. Signal strength at the service area contour may be less than 37 dBu.

[3] When the actual antenna HAAT is greater than the reference HAAT, the allowable ERP will be reduced in accordance with the following equation:

$$ERP_{allow} = ERP_{max} \times (HAAT_{ref} / HAAT_{actual})^2.$$

[4] Applications for this service area radius may be granted upon specific requests with justification and must include a technical demonstration that the signal strength at the edge of the service area does not exceed 37 dBu.

TABLE 4 450–470-MHz ERP/HAAT limitations

450 to 470 MHz - Maximum ERP/Reference HAAT for a Specific Service Area Radius										
Service area radius (km)	3	8	13	16	24	32	40[4]	48[4]	64[4]	80[4]
Maximum ERP (w)[1]	2	100	500[2]	500[2]	500[2]	500[2]	500[2]	500[2]	500[2]	500[2]
Up to reference HAAT (m)[3]	15	15	15	27	63	125	250	410	950	2700

[1] Maximum ERP indicated provides for a 39-dBu signal strength at the edge of the service area per FCC Report R-6602, Fig. 29 (see § 73.699, Fig. 10).

[2] Maximum ERP of 500 watts is allowed. Signal strength at the service area contour may be less than 39 dBu.

[3] When the actual antenna HAAT is greater than the reference HAAT, the allowable ERP will be reduced in accordance with the following equation:

$$ERP_{allow} = ERP_{max} \times (HAAT_{ref} / HAAT_{actual})^2.$$

[4] Applications for this service area radius may be granted upon specific requests with justification and must include a technical demonstration that the signal strength at the edge of the service area does not exceed 39 dBu.

TRANSIT RADIO SYSTEMS—USAGE AND PLANS

Survey Purpose and Approach

A survey of a representative sample of transit systems was undertaken to determine the status of transit communication systems and applications and to understand plans for future systems and applications. The survey collected information to aid in assessing the effects of FCC refarming.

The research team's approach to this information collection was as follows:

- Identify and Screen Interviewees—The research team

identified a balanced group of survey participants from the following categories:

- Type of transit system (e.g., bus, rail, or paratransit or taxicab),
- Location of system (i.e., urban, suburban, or rural), and
- Size of operation (i.e., large, medium, or small).

• Create Detailed Survey Questionnaire — This questionnaire served as the primary guide for in-person and phone interviews. The questionnaire and cover letter are provided in Appendixes A and B, respectively.
• Conduct Survey — The research team distributed the survey to most prescreened participants and scheduled an agreeable time for the detailed discussion. All but one interview were conducted by telephone. Part of the Massachusetts Bay Transportation Authority (MBTA) interview in Boston was conducted in person. Telephone interviews, as opposed to a mailed survey, proved essential because many organizations had multiple points of contact and many required further clarification.
• Summarize Results — The results of the survey were synthesized into a summary format and reviewed for inconsistencies. The research team conducted follow-up phone discussions to clarify any inconsistencies.

Survey Participants

Twenty-one transit operations participated in the survey. The research team was successful in contacting 17 of the 21 transit systems listed in the original workplan. Initial interviews indicated a significant percentage of transit systems were moving to 800/900-MHz radio systems. Although this is useful information, the research team felt that a broad understanding of the systems that will remain in the VHF and UHF bands was essential. The Southeastern Pennsylvania Transportation Authority (SEPTA) and the Chicago Transit Authority (CTA) were, therefore, added to the list. The Topeka, Kansas, Metropolitan Transit Authority (MTA) and the Denver, Colorado, Rapid Transit District (RTD) were also added to the survey. Table 5 lists the transit systems interviewed and provides a summary of the operation and location of each.

Radio System Characteristics

The transit operations interviewed have widely divergent radio systems with varying configurations and complexity. Nevertheless, they share several operational characteristics. Most systems are exclusive use, multichannel, nontrunked or simulcast, dispatch to mobile, analog technology for predominantly voice communication.

TABLE 5 Transit operations summary

Transit System	Vehicle Types and Quantity						Population Category	Region
		Rail						
	Commuter	Heavy	Light	Bus	Paratransit	Other		
Boston - MBTA	22	60	30	1000	subcontract	>400	Urban, Suburban	East
New York - NYCT		225		3700	0	4500 police	Urban, Suburban	East
Miami - Tri-Rail	30						Suburban	East
Washington - WMATA		108		1450		375	Urban, Suburban	East
Los Angeles - LACMTA	54			2000	subcontract	300+	Urban, Suburban	West
San Francisco - BART		39					Urban, Suburban	West
Chicago - CTA, METRA	50	108		2100	subcontract		Urban, Suburban	Central
Philadelphia - SEPTA	75	50	20	1200		300 maint	Urban, Suburban	East
St. Louis - Bi-State			31	650	75	105	Suburban, Rural	Central
Spokane - STA				144	65	25	Rural	West
Fresno - FTS				107	20	5+	Rural	West
Salt Lake City - UTA				492	25	50	Suburban	Central
Denver - RTD			11	860	70	150	Urban, Suburban	Central
New Orleans - RTA			41	477	30		Suburban	Central
Boise Urban Stages				27	4	5	Rural	Central
Tacoma - Pierce				180	105	29		West
Topeka - MTA				33	10	10	Rural	Central
Denver - Metro Taxi						350	Suburban	Central
Atlanta - Checker Cab						200	Urban, Suburban	East
Evansville - River City Cab						38	Rural	Central
Pomona - Diversified Paratransit						55	Rural	West

Note
1. For responses given in number of train cars, assume 8 cars average per train

The systems are used for voice communications largely under exception (i.e., emergency, maintenance) conditions or for yard management. Ten systems dedicate one or more channels to data, mostly for bus and paratransit routing purposes. San Francisco's Bay Area Rapid Transit (BART) is the sole system significantly using simulcast. The radio systems are important for normal operation of rail and bus operations and are critical to the operation of paratransit and taxi services. Some systems use radio signalling systems for rail. System coverage radius ranged from 5 to 50 mi.

The three major frequency bands used are 150- to 174-, 450-to 470-, and 800/900 MHz. Lowband VHF (< 60 MHz) and 502 MHz are used for small portions of some systems. Seven of the interviewees use 800/900-MHz trunked or non-trunked systems, with five others planning 800/900-MHz systems within 2 years. All radio systems are analog with the exception of the recently commissioned Denver Transit Radio system.

Radio Component Characteristics

The components of transit radio systems typically consist of mobile and portable terminal equipment, base stations, microwave or leased line backbone, and control centers. Mobiles or portables are found on essentially all revenue and nonrevenue vehicles. Mobiles constitute most of the installed base. New purchases lean toward portables because of the flexibility to assign a portable to each operator or other staff. Most terminal equipment has emergency alarm capability; some has automatic number identification.

Most systems operate base stations at less than 300 watts ERP and do not consider their radio coverage to exceed their operating area. Many interviewees had difficulty providing ERP and HAAT information.

For the systems surveyed, Motorola and General Electric (now Ericsson) represent approximately 80 percent of the installed equipment base. Other vendors identified include E.F. Johnson, Midland, Bendix King, and Yaesu. The research team obtained a good cross section of system age, ranging from less than 1 year to more than 20 years. Eight transit operators had radio systems or major subsystems less than 5 years old.

Medium to large transit operations use cellular phones and pagers extensively for supervisory and maintenance personnel. CTA and the Washington Metropolitan Area Transit Authority (WMATA) paging systems are integral to their radio systems.

Tables 6 and 7 summarize the system-and component-level characteristics of the 21 interviewee systems.

System Use

System resources for multimode transit operations are typically allocated on a functional basis. Rail, bus, paratransit/transit, security, and maintenance have dedicated channels or subsystems in most transit systems. The functional system partitions are further divided by regions for large operators. For example, the New York Metropolitan Transit Authority (NYMTA) has a dedicated five-frequency-pair VHF subsystem for MTA police. The five pairs are divided among the four major New York boroughs, with one citywide frequency pair. Channel loading is low to medium for normal conditions and medium to heavy for exception conditions. Police and supervisory channels represent the most use during normal conditions.

Only commuter rail operations share facilities and radio frequencies with other rail operations (e.g., Conrail and Amtrak). Boston's MBTA has eight channels in the 160-MHz band and shares three of the eight with Amtrak and others. Commuter rail operations are most limited in their options relative to refarming and rely heavily on AAR to represent their interests to the FCC.

Most systems have no compatibility with local fire or police department radio systems. Three interviewees expressed a desire for future compatibility. The Los Angeles County Metropolitan Transit Authority (LACMTA) shares underground leaky coaxial antenna facilities with local police and fire departments. BART owns a 160-MHz system for fire department use in its tunnels. On the other hand, St. Louis firefighters drive a communications van into the area for tunnel fire situations.

System Performance

The interviewee systems generally considered the performance of their radio systems to be acceptable. Several indicated they had no real problems beyond minor dead spots under normal conditions, and interference was rarely mentioned. Only Chicago and New York mentioned frequent interference, coverage, and congestion problems. Essentially all systems experience congestion during exception conditions. Obsolescence was mentioned by several operators planning near-term radio system upgrades.

Table 8 summarizes radio system usage and performance data collected during the interviews.

Plans and Investments

Many transit systems contacted are planning major investments in their radio infrastructure. Thirteen plan major upgrades or new systems within the next 5 years. Budget estimates range from $150,000 for a new trunked Boise, Idaho, system to $40 million each for BART and SEPTA. Financing is typically federally supported at 50 to 80 percent for transit authorities and by revenues for taxi services. Obsolescence and congestion are the primary reasons for these investments.

TABLE 6 Radio systems characteristics 1

Transit System	Vehicle Type	Band (MHz)	No. Channels	Shared?	Simulcast/ Trunked	ChannelUsage	Other Features
Boston - MBTA	Hvy/Lgt Rail	470	11	N	N	voice	
	Commuter	160	8	Y, 3	N	voice	
	Bus	470	2	N	N	voice	
	Police	470	2	N	N	voice	
	Bus	30	2	N	N	voice - voted	
	Paratransit	470	1	N	N	voice - voted	
	Maint	44	2	N	N	voice - simplex, voted	
New York - NYCT	Police	VHF hi	5pr	N	N	voice - 4 boroughs, 1 city	above/below ground
	Heavy Rail	VHF	3.5pr	N	N	voice - simpl/duplx	above/below ground
	Bus/super	800	20	N	T	voice, minor data	
	Comm Rptr	470	4pr	N	N	voice	
	Maintenance	VHF	1	N	N	voice - simplex	
	Revenue	SMR	rent	Y	N	voice	
	Heavy Rail	470	1 pr	N	N	voice	
	Track	UHF	1pr	N	N	voice	
	Police	SMR	-	Y	-	voice	
Miami - Tri-Rail	Signalling	900	1	Y	N	data only	
	Maint/yard	161	2	N	N	voice	
	Rail	161	1	Y	N	voice	
Washington - WMATA	Rail	160	7	N	N	voice	ANI
	Bus	450	8	N	N	7 voice, 1 data	ANI
Los Angeles - LACMTA	Bus	470	15	N	N	9 voice, 6 data	AVL, polled, 2 min. @ 2400baud

Five interviewees are moving to 800/900-MHz systems, in part to avoid the refarming issue. Half of the transit operations contacted will be operating at 800 or 900 MHz within a few years. Analog systems are not being replaced yet by digital systems. Only one operator definitely plans a move to a digital system. Two others are considering digital but are dissuaded by the cost. Price and product availability remain the major reasons for staying with analog.

System operators are not planning major investments in advanced features. AVL and low baud rate data capability is planned for most new systems. These AVL systems operate in a polled or alarm mode and do not require high baud rates. A communication space of 2400 baud is considered acceptable for this function. Interviewees foresee advanced data requirements for real-time bus location, fare box monitoring, and remote diagnostics eventually. Only CTA mentioned ITS or other future system requirements.

The new 800-MHz BART system used its right of way effectively to minimize its radio system investment. The contractor was allowed to install multifiber optical cable along the right of way as a backbone for the new BART system. The contractor has the right to sell the use of excess capacity for "information highway" and telecommunication services. Revenues from this venture will be credited against the $40 million upfront investment by BART.

Table 9 summarizes survey participants' plans and investments.

Perspective on Refarming

The awareness and understanding of the FCC refarming activity were discussed during each interview. The awareness level is generally proportional to the size of the operation. SEPTA and LACMTA were very aware of refarming, whereas most taxi services had little or no knowledge of refarming activities. The small transit and taxi systems rely heavily on their equipment vendors for information of this type. Some appear to believe—incorrectly—that their existing or new system will not be affected or could be modified via software.

Awareness of options to respond to refarming varied, although not necessarily by size of operation. The options mentioned were moving to 800/900 MHz and purchasing new terminal and base station equipment. The staff contacted were asked to estimate the cost effect of refarming if they were required to meet the narrowband requirements. Many survey participants could not immediately estimate the replacement cost for their equipment, except for those planning system replacement in the near term. Thus the cost effect is not well quantified.

TABLE 6 Radio systems characteristics 1 (continued)

Transit System	Vehicle Type	Band (MHz)	No. Channels	Shared?	Simulcast/ Trunked	ChannelUsage	Other Features
	Super/Police	450	4	N	N	voice	
	Subway	160	6	N	N	2pr voice, 1 pr data	
	Tow Trucks	900	7	N	T	voice	
	Surface Rail	900	5	N	T	voice	
San Francisco - BART	Maint	43	1	N	N	voice - simplex	
	Train Control	160	1	N	S	voice - duplex	
	Admin	160	1	N	S	voice - simplex	
	BART Police	453	1	N	S	voice - duplex	
	BART Police	450	1	N	S	voice - duplex	
	S.F. Fire	160	1	N	N	voice - duplex	
Chicago - CTA, METRA	Bus	450	8	N	N	6ch voice, 2 ch data	data is AVL
	Rail	470	8	Y, 1	N	voice	
	Commuter	160	14	Y, 7	N	voice - simplex, 1 police, 6 districts	
Philadelphia - SEPTA	Commuter	160	7	Y, 3	N	voice - 2 road, 1 maint, 1 yard	
	Bus &	502	13	N	N	voice - 1 police, 1 sub, 11 bus	
	Hvy/Lt Rail						
	Light Rail	30	2	N	N	voice	
St.Louis - Bi-State	Light Rail	800	5	N	N	voice - duplex ,3 garage, maint/super, paratransit	
	Bus & Para	458	5	N	N	voice-duplex, ops, maint, yard, security, super	
Spokane - STA	Bus & Para	450	5	N	N	voice	

TABLE 6 Radio systems characteristics 1 (continued)

Transit System	Vehicle Type	Band (MHz)	No. Channels	Shared?	Simulcast/ Trunked	ChannelUsage	Other Features
Fresno - FTS	Bus & Para	450	2	N	N	1 Voice, 1 Data	
Salt Lake City - UTA	Bus & Para	450	5	N	N	5 Voice & Data	Telemetry
Denver - RTD	Bus & Lt Rail	450	10	N	N	2 Data, 7 Voice, 1 Unused	AVL
New Orleans - RTA	Bus, Para, LR	800	10	N	N	9 Voice, 1 Data	
Boise Urban Stages	Bus & Para	450	2	Y, 1	N	voice	
Tacoma - Pierce	Bus & Para	900	5	N	N	voice	
Topeka - MTA	Bus & Para	450	2	N	N	voice	
Denver - Metro Taxi	Taxi	450	2	N	N	voice	
		800	1	N	N	data	
Atlanta - Checker Cab	Taxi	450	2	N	N	1 voice, 1 data	
Evansville - River City Cab	Taxi	150	2	N	N	voice	
Pomona - Diversified Paratransit	Taxi	150	2	N	N	1 voice, 1 data	

Notes

1. Quantities for radio equipment are approximate

2. S= simulcast

3. Essentially all have ANI capability that is rarely used

4. All systems are analog except Denver and Utah

TABLE 7 Radio systems characteristics 2

Transit System	Vehicle Type	Bands (MHz)	Mobiles	Portables	Base Number	Station ERP	HAAT	Coverage Radius (mi)	Average Age (yr)	Manufacturer
Boston - MBTA	Hvy/Lgt Rail	470	680	205	24 (20)	~90	<100	15		
	Commuter	160	500	307	11	75	300-1100	30	20	Motorola, Uniden
	Bus	470	1000	0	2	60-150	350-450	15		Motorola, GE
	Police	470	60	100	27(4)	~90	<100			Motorola
	Bus	30	400	300	5	100	350-450	15		GE
	Paratransit	470	60	0	2	250	350-450	10		Motorola
	Maint	44	100	140	2	100	650	30		Motorola
New York - NYCT	Police	VHF hi	65	4500	131 (46)	60	60-1k	30	15	GE Master II
	Heavy Rail	VHF	1800	2000	13	60			10	GE, Motorola
	Bus/super	800	3700	350		100			4	Motorola
	Comm Rptr	470	300	400		60				Motorola
	Maintenance	VHF	10	4		60				GE
	Revenue	SMR	90		0					Motorola
	Heavy Rail	470	110	50	1	60				GE/Mot
	Track	UHF	12	9	9	60				
Miami - Tri-Rail	Signalling	900	30		4	35-136	~200	~30	<2	Motorola
	Maint/yard	161			3	~35	50-100		10	Motorola
	Rail	161	30		5	35-136	50-100	~30	10	Motorola
Washington - WMATA	Rail	160	1300	700	41 (5)	40	100-600	30	10	Motorola
	Bus	450	1600	700	4	100	100-600	30	20	Motorola, GE
Los Angeles - LACMTA	Bus	470	2000		15	>300	1k-3k	50	15	GE
	Super/Police	450		~400	8	>300	1k-3k	50	10	GE, Mot
	Subway	160	100	300	5	10			<3	Motorola
	Tow Trucks	900							<1	Motorola
	Surface Rail	900	200	350	2	300	2000		<1	EF Johnson
San Francisco - BART	Maint	43	600	600	29 (19)	100	3000	?	>10	GE Master 2E, no breakdown by band
	Train Control	160								
	Admin	160								
	BART Police	453								
	BART Police	450								
	Fire	160								

Some of the comments provided by interviewees were as follows:

- "I cannot be sure my planned investment will not be wasted without an FCC ruling."
- "Eight hundred or 900 MHz is not an option, as no frequencies are available in this area."
- "I don't think our data needs can be met with 6.25-KHz channels."
- "Our vendor did not mention refarming when the research team bought our new system."
- "We have not thought about refarming at all."
- "How can I (confidently) work with my vendor to minimize the effect?"

Table 10 summarizes the refarming perspectives of survey participants.

TRANSIT COMMUNICATIONS APPLICATIONS AND TECHNOLOGY

Introduction

Transit agencies increasingly employ new technology to improve the safety, performance, and use of their transit systems. Radio communications are essential to supporting the daily operation of these transit systems. The ongoing FCC refarming activity will influence the direction of radio communications as the research team knows it today, which may affect agencies' attempts to move forward with operational improvements. This section identifies current and future transit applications and supporting communications technologies, particularly in light of the refarming initiative.

This section provides a brief overview of the following:

- Current and emerging transit system applications;
- Existing communications technology, new communications technology, and other related technology to support communications applications;
- Standards activities influencing technology direction; and
- Product direction of the major vendors.

Current and Emerging Transit System Applications

Current transit systems applications that rely on radio communications have remained essentially unchanged for several decades. Traditional voice communication via the

TABLE 7 Radio systems characteristics 2 (continued)

Transit System	Vehicle Type	Bands (MHz)	Mobiles	Portables	Base Number	Station ERP	HAAT	Coverage Radius (mi)	Average Age (yr)	Manufacturer
Chicago - CTA	Bus	450	2100		1+3rcv	140 & 750	637	30	>15	Motorola Metrocom
	Rail	470		1450	1+23rcv		637	30	>15	Motorola
- METRA	Commuter	160	600	450	60 (42)	200	50-200	30	<10	Motorola
Philadelphia - SEPTA	Commuter	160		600	47 (7)	60	100-500	35	12	Motorola
	Bus &	502	1200	600		20-30			15	Motorola, GE
	Hvy/Lt Rail									
	Light Rail	30		100						
St.Louis - Bi-State	Light Rail	800	750	105	9	200	350	20	3	Motorola, GE
	Bus	458			6	500	350	20	>15	Motorola, GE
Spokane - STA	Bus & Para	450	234	27	1		2200	10	>7	Motorola, Visor
Fresno - FTS	Bus & Para	450	120	26	2	75 watts			>8	Motorola, GE
Salt Lake City - UTA	Bus & Para	450	567	55	7	200,800	4500-8500	28		GE, Midland, King
Denver - RTD	Bus & Lt Rail	450	1200	139	22	75	1000-3000	27	2	Motorola
New Orleans - RTA	Bus, Para, LR	800	477	120				13	>10	Motorola
Boise Urban Stages	Bus & Para	450	39	5	2	100			>10	Motorola, GE
Tacoma - Pierce	Bus & Para	900	230	93	2	30	8	12	3	Motorola
Topeka - MTA	Bus & Para	450	53	6	1			4	20	Motorola
Denver - Metro Taxi	Taxi	450	350	10	2	300	5370	4	10-12	Motorola, Midland
		800	350	0	2	155	5730		8-10	
Atlanta - Checker Cab	Taxi	450	250	1	2	100	150	4	5	Motorola
Evansville - River City Cab	Taxi	150	38	0	2	75	125	4	7	Motorola
Pomona - Diversified Paratransit	Taxi	150	55	2	2	75	100	11	<1	Uniden, Motorola

Notes
1. HAAT data are estimates of highest or range of tower heights
2. Base station column format is total number (number above ground) (i.e. 25 (20))
3. Dominant manufacturer listed, most have a mix of manufacturers
4. Quantities are approximate

radio system provides a medium for vehicle operations, maintenance, management, and security personnel to receive instructions and provide status. Computer-aided dispatch (CAD) and AVL applications were added to some systems starting in the 1970s. CAD uses data communications and a video display and keyboard for vehicle dispatch and receipt of status reports.

AVL is a data application applied in emergency situations to pinpoint vehicle location. The existing AVL systems typically use RF signposts to establish a vehicle's location. All these voice and data applications use the traditional macrocellular radio system architecture of a control center, one or more base stations, and mobile terminal devices.

Current voice and data dispatch and reporting systems focus on "person-to-person" communications between transit system personnel. Future applications will integrate computing, communication, and navigation technologies to create "computer-to-computer" and "computer-to-customer/operator" communication, in real time.

Table 11 summarizes current and future transit system applications and their operational usage. This table is derived from information gathered during the transit system survey conducted by the research team. The information reflects the near-to medium-term expectations of a representative sample of transit operations.

The survey results indicate that this group is focused on traditional and "natural" extension applications for its radio systems. Most future applications are to be quasi-real-time data applications.

A more visionary view of future transit system applications is provided by the APTS program sponsored by the Federal Transit Administration (FTA). APTS, part of the U.S. DOT ITS program, is to develop systems that improve the public transit options available to passengers. The APTS program has partitioned potential applications into the following three categories:

- Smart Traveler — Applications that allow travelers to make immediate, informed decisions about their travel modes or routes, at any point in the process;
- Smart Vehicle — Applications that improve vehicle and fleet planning, scheduling, and operations in order to increase reliability, efficiency, and safety; and
- Smart Intermodal — Applications that provide information links between the various modes of public transportation and other ITS applications.

TABLE 8 Radio systems usage and performance

	Channel Utilization		Compatibility Needs			Problems (under normal conditions)				
Transit System	**Normal**	**Exception**	**Police**	**Fire**	**Other**	**Coverage**	**Interference**	**Congestion**	**Obsolescense**	**Other**
Boston - MBTA	Medium	Heavy	N	N	AAR		O - commuter	F - police, hvy rail	Y	
New York - NYCT	Medium	Heavy	desired	desired	Emer		O - police,rail	F - Bronx	Y	
Miami - Tri- Rail	Low	Medium	N	N	AAR		O - lightning			
Washington - WMATA	Low	Medium	desired	desired		F - bus				
Los Angeles - LACMTA	Low	Heavy	patch	patch		O- bus				Theft
San Francisco - BART	Med - police	Heavy	N	on 160MHz		F - police			Y	
Chicago - CTA, METRA	Medium	Heavy	desired	desired	AAR	F - all		F - bus,rail	Y	
Philadelphia - SEPTA	Medium	Heavy	desired		AAR	F - bus,rail			Y	
St. Louis - Bi-State	Low	Heavy	N	N			O		Y - 450MHz	
Spokane - STA	Low	Medium	N	N						
Fresno - FTS	Medium	Heavy	N	N		O	R - truckers			
Salt Lake City - UTA	Medium	Heavy	N	N			R			
Denver - RTD	Medium	Medium	N	N		R		O		
New Orleans - RTA	Medium	Heavy	N	N		R		O	Y	
Boise Urban Stages	Low	Medium	N	N					Y	
Tacoma - Pierce	Medium	Heavy	N	N				R		Mix D & V
Topeka- MTA	Low	Medium	N	N			R		Y	
Denver - Metro Taxi	Heavy	Heavy	N	N		O	R	O		
Atlanta - Checker Cab	Heavy	Medium	N	N			R			
Evansville - River City Cab	Low	Medium	N	N						
Pomona - Diversified Paratransit	Medium	Heavy	N	N		R	F	R		

KEY: R = rare O = occasional F= frequent Y= Yes N= No

Notes

1. Patch indicates connected through a control center

2. Problems legend

3. All experience congestion during exception conditions

rev C 4/6/95

Table 12 lists the potential applications within each of the three categories.

Current and Emerging Communications Technologies

Transit-related communications technologies are undergoing significant change. Current communications technologies are analog in nature and used largely for voice communication. Current architectures use transit-owned infrastructure (or shared in the case of commuter rail). Emerging communications technology, on the other hand, is digital and supports higher speed data along with voice. Transit-owned infrastructure is being supplemented by other public or private services, such as cellular telephone and Global Positioning System (GPS) satellite transmissions for AVL.

Beyond the transition to digital technology, emerging technology includes new terminal devices and enhanced software to support applications that integrate stand-alone functions. For example, in an AVL application, GPS receivers on vehicles allow geographic coordinate information to be fed to the central control center, which uses software-based mapping and information systems to determine the street location. The underlying radio system has not changed—its use has changed.

Table 13 summarizes the current and emerging technologies used to support the applications identified in the transit survey. The traditional applications of voice communication and CAD may soon employ digital, high-speed technology. The new applications—traveler information systems, automated maintenance, and electronic fare collection—all integrate existing (but emerging for the transit industry) technology. GPS appears to be the location aid of choice by transit and numerous other industries.

On the basis of the results of the transit survey conducted by the research team, the transition to digital radio technology will be gradual. The transit agencies that the research team contacted do not yet consider the advantages of digital radio as relevant to their requirements.

APTS Technology

Under the APTS program, it is acknowledged that most, if not all, of the technologies needed to implement APTS exist. The challenge is to integrate these technologies effectively and efficiently into valued applications. Most of these technologies are not communications technologies; how-

TABLE 9 Plans and investments

Present System		Upgrade Plans					
Transit System	**Replacement Cost/Investment**	**What?**	**Why?**	**Band/Channels**	**Advanced Technology**	**Budget**	**Timing**
Boston - MBTA	$10M invested	all but commuter	congestion	800MHz, 20ch	AVL, trunked	$25M	1997
New York - NYCT	Unknown	Police	congestion	470MHz, 15ch	digital, data	$30M	1996
		Misc	congestion	470MHz trunked	TBD	TBD	2000
Miami - Tri-Rail		Signalling	ATCS capability	900MHz	more data - AVL		1996
		ARDIS services	Train work orders				>2000
Washington - WMATA	$5M bus	police	obsolete, congestion	490 or 800MHz trunked	analog or digital, MDT	$5M	1996
Los Angeles - LACMTA	$3-10M replace rail		rail add-ons	900MHz	analog only	$12M	1998
	invested $20M bus		rail-add-ons	160MHz			
San Francisco - BART	$20M	whole system	obsolete, performance	800 MHz trunk	analog, data capable	$40M	1995
Chicago - CTA	Uncertain	Bus	obsolete, congestion	450MHz, trunked	AVL	confidential	1995
- METRA	$5-7M	none					
Philadelphia - SEPTA	$20M bus in 1980	Bus/Rail	obsolete, refarming	502MHz, 12.5kHz digital, & 900 MHz	AVL, bus data,	~$40M	1998
St. Louis - Bi-State	$2M	paratransit	data dispatch needs	450Mhz new chnl	mobile data terminals		1995
		bus	obsolete,rail compatible	800Mhz trunked	analog or digital, 5 ch	$5M	1996
Spokane - STA	$0.7M	Mobile	Replacement	450			1997
		AVL	Upgrade	450			2005
Fresno - FTS	$0.5M	New System	Age	450 MHz/2 Ch.	Analog only	800000	1997
		AVL	Upgrade	450 MHz/2 Ch			2005
Salt Lake City - UTA	$2M	Bus	Efficiency, Refarming	450 MHz/5 Ch	Digital only	375000	2000
Denver - RTD	$11M	Bus & Para	Growth	450 MHz/10 Ch	Digital, AVL	$1M	1997
New Orleans - RTA	$1.5M	New system	Congestion	800 MHz/10 Ch	Trunking	700000	1997
Boise - Urban Stages	Unknown	New system	Efficiency, refarming	800 MHz/2 Ch	Trunking	150000	1997
Tacoma - Pierce	$3M		Expand base stations	900 MHz/5 Ch			1997
Topeka - MTA	Unknown	New System	Replacement	Unknown	Unknown	150000	1997
Denver - Metro Taxi	$0.5M invested	New System	Replacement	450 MHz/2 Ch	Analog only	100000	1997
Atlanta - Checker Cab	$0.7M	AVL	Upgrade	450 MHz/4 Ch	Analog only	100000	2000
Evansville - River City Cab	$40k		add auot-dispatch	150 MHz/2 Ch		80000	1996
Pomona - Diversified Paratransit	$60k	Base Station	Replacement	150 MHz/2 Ch	Analog only	10000	2005
Notes							
1. Quantities are approximate			*3. Taxi financing by revenue*				
2. Transit authority financing via Federal "Section 9" for 50-80% reimbursement							

ever, they require a communications infrastructure and local access to a communications terminal device. Table 14 lists the APTS available technologies by category.

The use of these APTS technologies will significantly affect the use and overall spectrum needs of the transit communications systems. The ITS America Telecommunications Committee and TRB Committee on Communications are working to define the spectrum needs for IVHS applications, including those needs specific to transit. The *Transportation Research Circular* article of June 1994, "Spectrum Needs for Intelligent Vehicle Highway System Application," provides matrices detailing IVHS applications, possible technologies, required RF spectrum, and RF power levels.

Underlying Technology of Radio Systems

Improvements in the underlying technologies for radio systems enable advanced features and performance improvements while decreasing the equipment size, weight, and cost. This document has grouped these technologies under digital radio technology and component technology. The trends for each of these groupings are discussed in the following paragraphs.

Digital Radio Technology

Although the major radio manufacturers offer narrowband analog radios that comply with the FCC's first proposed level of spectrum efficiency requirements under refarming, these radios are essentially modified versions of the same radio technology that has existed for years. The recent technological advances are predominantly related to the new digital radio equipment; therefore, the research team will concentrate on those advances.

Technology, as it applies to the performance of digital PLMR systems, can be categorized as follows:

- Modulation schemes,
- Channel access schemes,
- Voice coding schemes,
- Error correction,
- Encryption, and
- Data transmission.

TABLE 10 Refarming perspectives

Transit System	Known Alternatives	Costs/Impact	Comments/Concerns
Boston - MBTA	move to 800MHz	$25M	Commuter rail cannot move to 800 MHz & has larger service area. 6.25kHz data is an issue
New York - NYCT	limited 800MHz available clear migration path essential	Uncertain	No 800/900MHz left. Need >100w ERP. Total redesign-replace antennas and base stations
Miami - Tri-Rail	900MHz for data use ARDIS type services		
Washington - WMATA	move to 800MHz	Uncertain	Need interoperability
Los Angeles - LACMTA	move to 800/900MHz is partial solution	Uncertain - substantial	Cost mostly in Handhelds. Need resolution to allow planning; 12.5khz for digital is an issue
San Francisco - BART	move to 800 MHz	$40M	Releasing PLMR frequencies
Chicago - CTA	Uncertain	Uncertain	looking at obtaining added (refarmed) channels
- METRA	Upgrade	$5-7M	How to work with vendor to minimize impact
Philadelphia - SEPTA	UHF digital, 900MHz	$40M	
St. Louis - Bi-State	use refarm to justify 800MHz	Uncertain	
Spokane - STA	New 450 MHz system	Uncertain - substantial	Pretty happy with current system. Have not thought about FCC refarming issues.
Fresno - FTS	Equipment Upgrade	Uncertain - substantial	Unconcerned. No plans at this time.
Salt Lake City - UTA	Digital System	100,000	Current system takes refarming into account. System ready to be upgraded to 100% digital.
Denver - RTD	System replacement	$11 M	Never discussed refarming issues. Does not know if new system can easily adapt to FCC rule changes.
New Orleans - RTA	Not Applicable		
Boise Urban Stages	800 Trunking	150,000	New system will not be affected FCC rule changes. Cost 50% more than 450 MHZ system
Tacoma - Pierce	Not Applicable		

TABLE 10 Refarming perspectives (continued)

Transit System	Known Alternatives	Costs/Impact	Comments/Concerns
Topeka - MTA	Digital technology		Will consider digital if cost-effective. Will get new system, but no plans as to what. Not aware of refarming and is now concerned
Denver - Metro Taxi	New System	1,000,000	Not interested in cellular or trunking. Wants to stay in the lower frequencies
Atlanta - Checker Cab	New System	$1-2,000,000	Will cause hardship
Evansville - River City Cab	New System	50,000	Old equipment has no resale value. Concerned about clarity. Will not go cellular.
Pomona - Diversified Paratransit	New System	100,000	Hopefully, new digital dispatch system will work with narrow band.

Modulation Schemes. Modulation schemes relate to the spectral efficiency, which is measured in the number of bits-per-hertz transmitted. The modulation scheme can also affect the range of system coverage by affecting the transmitted power (measured in energy per bit or per symbol) and the receiver threshold at which point the bit-error ratio becomes intolerable. Furthermore, the choice of modulation scheme can affect the transmitter efficiency (measured as the amount of power radiated relative to the total consumed power), which in turn affects the battery life of a portable.

Modern modulation schemes employed in portable units use constant-envelope modulation that employs efficient nonlinear amplifiers. At the same time, base transmitters will use less efficient linear amplifiers and modulation schemes that optimize the receiver of the mobile terminal.

Channel Access Schemes. A channel access scheme enables the system to share spectrum among separate connections or circuits. The channel access scheme also is related to the channel bandwidth because each channel access scheme is optimized to operate in a different channel plan.

Narrowband or very-narrow narrowband channel frequency division multiple access (FDMA), for example, re-

TABLE 11 Current and future system applications

System Application	Current Use	Future Use
Voice Communications	- Emergency - Status - Dispatch	- Emergency - Status - Dispatch - Telephone Interconnect
Computer-Aided Dispatch (CAD)	- Dispatch - Reporting	- Dispatch - Reporting - Graphics (i.e., maps)
Automatic Vehicle Location, Tracking & Guidance	- Emergency Location	- Emergency Location - Tracking/scheduling
Traveler Information Systems	- Trials Only	- Real-time Vehicle Status to Current or Potential User - Database Information
Automated Maintenance	- Trials Only	- Real-time Health and Use Data Collection
Electronic Fare Collection	- Trials Only	- Debit Farecard Remotely

TABLE 12 APTS potential applications

Smart Traveler	Smart Vehicle	Smart Intermodal
- Real-time Information - Integrated Fare Payment - Home or Workplace Multimodal Information - Wayside and Onboard Traveler Information - Integrated Billing - Dynamic Ridesharing	- Auto-demand-responsive Dispatching - Onboard Auto-guidance - Fleet Monitoring, Control and Management - Real-time Data Collection and Analysis - Schedule Planning - Passenger Information - Equipment Performance Monitoring - Driver Performance Assistance - Driver and Customer Security	- Traffic Signal Timing & Priorities - Auto Toll Collection and HOV Verification - HOV Lane Access Control - HOV Toll Lanes - HOV Bypass Lanes at Metered Ramps - Centralized Transportation Management - Real-time Intermodal Travel Information-integrated Fare/toll Parking Fees

Note: HOV = high occupancy vehicle

fers to the proposed split channels of 12.5 KHz or 6.25 KHz. TDMA refers to a 25-KHz channel with four or six time slots or a 12.5-KHz channel with two or three time slots. Code division multiple access (CDMA) requires a considerably larger channel that supports many circuits or voice conversations.

Narrowband FDMA with channel splitting is the spectrum efficiency approach chosen by the FCC in the refarming proposal, adopted by the federal government for all federal channels, and specified by APCO Project 25. A disadvantage to narrowband FDMA is the requirement for very precise, sharp filters and tighter transmitter frequency control (to keep the transmitter in the center of the assigned channel, thus mitigating adjacent channel interference). FDMA equipment for 12.5-KHz channelization is offered by several equipment suppliers.

When TDMA is applied to refarming, two users, assigned to the same channel, will each have use of 100 percent of the channel bandwidth 50 percent of the time (the time slot duration is in the millisecond range so that the channel is switched between users very rapidly and is undetectable). FDMA with channel splitting, on the other hand, will pro-

TABLE 13 Current and emerging communications technology

System Application	Current Technology	Emerging Technology
Voice Communications	- Conventional Analog - 15/25-KHz Channels	- Trunked Digital - 12.5-KHz Channels
Computer-Aided Dispatch (CAD)	- Analog RF Modem - 15/25-KHz Channels - 2400 Baud	- Digital RF Modem - 12.5-KHz Channels - 9600 Baud or Higher
Automatic Vehicle Location	- RF Signpost	- GPS (Satellite) - RF Transponders
Traveler Information Systems		- Digital RF Modem - Integrates CAD, AVL, and remote displays
Automated Maintenance		- Digital RF Modem - Integrates CAD and On-vehicle Sensors
Electronic Fare Collection		- Digital RF Modem - Integrates On-vehicle Farebox and Customer Database

TABLE 14 APTS available technology

Smart Traveler	Smart Vehicle	Smart Intermodal
- Smart Cards - Computers and Telecommunications - Touch-tone Telephones - Voice Synthesis - Cable, TV, Radio - Interactive Video Displays - Roadside Monitors - Wayside and Onboard Displays - Smart Kiosks - Dynamic Multimodal Database	- Component Sensors - Automatic Passenger Counters - Automatic Vehicle Location - Data/voice/cellular Radio - Computer-Aided Dispatch - Smart Card Readers - Audiotex/videotex Displays - Geographic Information Systems	- Automatic Vehicle Identification - Image Processing - Vehicle Guidance Systems - Dynamic Multimodal Dispatching Software - Integrated Adaptive Signal Timing and Traffic Management Systems - Smart Card Systems

vide each user with 50 percent of the channel 100 percent of the time.

TDMA requires a common controller to synchronize time slots; therefore, unless a user can justify using all the time slots in a channel, the user must consolidate with others to share the common controller (which implies that all users have the same geographical coverage requirements) or the spectrum is wasted. (The channel access method called TDMA is actually an FDMA system with each channel further split into TDMA time slots.) At this time, TDMA is applied only to trunked systems. In the future, equipment may be offered that will enable TDMA to be used with conventional systems.

TDMA is inherently more spectrum efficient than channel splitting FDMA. That is because the narrowband or very-narrow narrowband FDMA channels require separation or guard band, which is wasted bandwidth. TDMA is capable of dynamic bandwidth allocation to data transmissions. Because the frequency plan used by a TDMA system uses wider channels than a comparable FDMA system, the data can potentially use the entire data throughput allocated to two or more voice channels, when those voice channels are idle.

The FCC specifically approved alternate plans with equivalent spectral efficiency in its R&O, thereby allowing the use of TDMA.

TDMA does not support "talk around" for mobile-to-mobile communications, requiring the use of either a non-TDMA-dedicated channel or a repeater, the latter requiring two voice channels. Also, TDMA is not backward compatible with today's FDMA radios, which presents some potential migration and evolution difficulties. At least one equipment supplier offers products that are reportedly compatible with both in a multimode operation.

CDMA (also known as spread spectrum) is a technology that evolved from military defense applications. By design, CDMA requires a wideband channel to operate. The existing and proposed FCC channel plans do not support this wideband channelization. (A detailed description of CDMA is beyond the scope of this document.)

Some cellular telephone operators and some of the emerging PCS operators are adopting CDMA as their channel access method of choice. In light of the benefits offered by CDMA, and with the emergence of new equipment to support cellular and PCS, CDMA may be used for PLMR in the future. CDMA is not being considered now. CDMA was considered and rejected by the APCO Project 25 Steering Committee because it was considered inappropriate for the public safety environment.

Voice Coding Schemes. To transmit voice (an analog signal by definition) over a digital radio requires a coder-decoder (CODEC) to digitize the analog signal. Non-compressed voice, however, requires high data speeds to be transmitted in real time, in the range of 256 Kbps. Companded voice requires 64 Kbps, and compressed Vocoded voice (represented by sound bits modeled after the human voice track, with redundant information removed) can be transmitted in 4 to 8 Kbps. The Vocoder specified by APCO Project 25 is being adopted in several products and other applications.

The choice of Vocoder determines the spectral density of voice circuits per channel and the naturalness of the human voice received. In addition, the extensive processing required to reduce the bit-rate of the voice is time consuming (resulting in speech delay) and power consuming (resulting in lower battery life).

Error Correction. Compressed digitized transmissions tend to multiply errors—a single bit-error in the transmission will result in many errors in the coded signal. For this reason, error-correction codes are applied to these systems to reduce the occurrence of otherwise routine errors to a small probability.

As a consequence of the error-correction, digital systems tend to work better in fringe areas up to a point. At that limitation point, they degrade very rapidly and cease to be functional. A user will find that a new digital system has clearer voice near the edge of the system coverage (relative to the old analog system) but will not work at all beyond the coverage area (whereas the old analog system worked sporadically and with noise and distortion).

Encryption. Although encryption does not affect radio system performance directly, it is related to the Vocoder and modulation of the radio. Most new digital radios offer encryption as standard or as an option to prohibit monitoring or eavesdropping on the channel. Encryption may be important in order to maintain the security of a transit system and to maintain confidentiality during sensitive situations.

Data Transmission. Digital radios convert voice into a digital data stream. Therefore, they are inherently suited for transmitting digital messaging or computer data. Most modern digital radios have a data rate of 9.6 Kbps, as opposed to the 2.4 Kbps typical of their analog counterparts. In addition, TDMA systems can use multiple time slots to achieve data speeds of 16 Kbps or higher.

Component Technology

Semiconductors. Semiconductor devices are progressing in three major areas: higher integration, lower voltages, and higher signal-frequency operation. Higher integration allows more functionality on a single chip, thereby decreasing size and weight and increasing the reliability of handheld, mobile, base station, and control center radios. Lower-voltage devices are becoming common for RF and digital devices in radio systems. Three volts or lower digital devices are available. Low-voltage operation translates into lower battery demands and longer operation between charges. Higher frequency devices are emerging through improvements in the traditional silicon materials used to make digital and RF devices. Newer developments in materials such as gallium arsenide and silicon germanium offer superior performance in the 800-MHz and above range and are becoming more cost-competitive.

Smart Antennas. Smart antennas, although not being used for PLMR, are being proposed for and applied to cellular telephone, commercial mobile radio (CMR), and PCS systems. A technology that has emerged from military radar phased array antennas, smart antennas can project dynamically changeable and adaptable radiation patterns that were previously unrealizable. The falling cost of gallium arsenide and digital signal processing elements has made it practical to apply this technology to commercial applications. If refarming includes a requirement for limitation of ERP as opposed to HAAT, some systems may benefit by applying smart antenna technology.

Batteries. Battery suppliers are struggling to satisfy the PLMR handheld requirements for 8- to 12-hour service as new power-hungry digital products emerge. Batteries account for nearly 50 percent of the weight of handheld radios. Today's handheld products typically use rechargeable nickel-cadmium (NiCd) batteries, and several offer high-capacity nickel metal hydride (NiMH) battery technology. NiMH offers approximately a 30 percent energy density increase over NiCd, does not suffer from a memory effect in which the usable capacity decreases, and is free of the environmental disposal concerns associated with NiCd.

Lithium ion is an emerging battery technology that provides a 200 to 300 percent increase in energy density as opposed to NiCD. Lithium ion is used in some portable computers, and radio products are just beginning to use this technology. Lithium-ion batteries will benefit transit systems by improving the operating performance and decreasing the size and weight of handhelds.

Wireless Wide Area Network (WAN) Technology

Public wireless WANs offer a viable supplement to dedicated transit radio systems and are worthy of brief mention. Wireless WANs now provide broad above-ground coverage throughout the United States. Transit systems can procure this "communications technology service" as needed for nonsafety or time-critical data transfers. Wireless WANs can serve to supplement existing capability or defer investment in new capacity.

There are two major suppliers of terrestrial wireless data services, and additional vendors are building systems. In addition, cellular phone service providers, which many transit operations already use, will soon offer data services known as cellular digital packet data (CDPD). These data services can support up to 19,200 baud with message response time of several seconds.

Low earth-orbiting and geosynchronous-orbiting satellites represent an emerging WAN technology for transit operations. GPS is only one of the satellite-based services available. Numerous transportation-related applications for satellite communications have been identified, including transmission of closed-circuit television (CCTV) traffic reports, dispatch, vehicle signing, traffic probe reports, traffic signal control, transit schedule reporting, weather and road condition reports, and incident reporting.

Satellites are another example of a communications infrastructure that transit systems can rent as needed. This service can supplement owned infrastructure in ways similar to that of cellular telephones at most transit operations.

CMR, formerly enhanced specialized mobile radio (ESMR), services are yet another WAN alternative for transit systems. These services require radios that operate in ways similar to cellular telephone, but they also offer dispatch, paging, messaging, data, and fleet services. There are companies that operate networks that provide good coverage for most metropolitan areas.

Related Land Mobile Radio Communications Standards

Standards are particularly important to transit agencies as technology progresses. Any review of emerging radio technology would be incomplete without discussion of standards efforts. The following section identifies the technical options and discusses APCO's efforts as well as other efforts to define standard architectures and communications methods.

The apparent user benefits that may be realized from the development of a standard are the following:

- Compatibility of equipment from multiple vendors and across vendors' product lines (to ensure that the user is not locked into a single-vendor solution for the service-life of the equipment),
- Interoperability between agencies and other, normally unrelated entities (e.g., between Transit and Public Safety during a disaster),
- Simplified radio equipment procurement,
- Competitive radio equipment procurement,
- Graceful system migration (forward and backward),
- Efficient use of the radio spectrum (assuming that is one of the objectives of the standardization),
- Mitigation or elimination of interference or other disturbance caused by other users in the same or adjacent channels,
- Clearly defined equipment performance benchmarks,
- Accelerated adoption of new technologies,
- Simplified and cost-reduced maintenance and testing, and
- Compliance with the FCC refarming initiative related to narrowband channelization (or other impending regulation).

Considering these benefits, it behooves the PLMR user to consider the impending standards efforts when attempting to define future PLMR communications needs.

APCO Project 25

APCO Project 25 is a joint government (i.e., local, state, and federal) and industry (i.e., both U.S. and international participation and interest) effort to develop technical standards for the next generation of public safety PLMR equipment. Specifically, the project relates to the use of state-of-the-art digital radio technology for both voice and data.

Project 25's goal is to develop a set of standards focused on six interfaces:

- Common air,
- Interconnect,
- Intersystem,
- Data,
- Host data, and
- Network management.

APCO Project 25 has a Memorandum of Understanding (MOU) with the Telecommunications Industry Association (TIA—the trade association of communications equipment manufacturers) and the telecommunications standards-setting branch of the American National Standards Institute (ANSI). Through this MOU and advisory group, TIA-25 was formed to support the initiative.

Project 25 was conceived in the hope that its recommendations would be accepted as the standard by TIA and ANSI; however, the mobile radio industry has not accepted this set of standards unilaterally. Although the effort has been widely commended, opponents point out that the Project 25 recommendations specify technology choices that are not optimum for all situations or applications. To date, TIA has adopted some portions of Project 25 as *Interim Standards* and some as *Telecommunications Systems Bulletins*.

Enhanced Digital Access Communications System (EDACS) Protocol Standardization

EDACS protocol is being submitted to TIA's TR-8 committee with a request to initiate a standard for Advanced Digital Land Mobile Radio. This protocol has been licensed to several PLMR equipment manufacturers.

Trans-European Trunked Radio Standard (TETRA)

In Europe, manufacturers representing the entire mobile radio community have been meeting for several years to develop standards for TETRA, employing four TDMA slots within a 25-KHz channel.

Vendor Products and Direction

Two vendors dominate sales of PLMR systems, but numerous U.S. and foreign vendors are striving to capture more of the market. Technical comparison of the various product offerings is the subject of much debate. The research team cannot hope to address this situation within the scope of this report but presents, in Table 15, a sampling of current and near-term technology available from one or more vendors.

TABLE 15 LMR technology status

Analog/ Digital	Conv/ Trunked	Channel Access	Channel Spacing	Availibility
Analog	Conv	FDMA	25 KHz	Now
Digital	Conv	FDMA	25/ 12.5 KHz	Now
Analog	Trunked	FDMA	12.5KHz	Now
Digital	Trunked	FDMA	12.5 KHz	late 1995
Digital	Trunked	F - TDMA	12.5KHZ	late 1995
Digital	Trunked	TDMA	25KHz	1995

No vendors contacted during this effort could offer a definitive migration path to 6.25-KHz systems, largely because of technical uncertainties.

Two vendors offer mobile radios supporting 6.25-KHz channelization in the 220-MHz band and claim easy adaption for UHF and VHF applications. These companies are mentioned by the FCC when users claim the technology is not available. Although the research team has not done detailed analysis, the research team does not believe UHF implementations will be easily achievable.

THE EFFECT OF SPECTRUM REFARMING ON TRANSIT

Overview

The research team has predicted the effects of refarming on transit operations on the basis of an understanding of (1) the FCC rules, (2) the vendor's current and planned products, and (3) the transit systems' current and future radio system usage and purchase plans. Many of the transit operations interviewed plan and need to make major radio equipment purchases before the end of this decade. Thus, knowledge of the potential effects of refarming is imperative for these operators to make rational choices.

Refarming's effect on a particular transit system varies depending on the degree of spectrum congestion in the service area. Congested areas are generally urban, while uncongested areas are generally rural. Most transit systems in congested areas are faced with insufficient channels to maintain an adequate grade of service (ability to communicate reliably with little or no delay), while systems in uncongested areas are essentially concerned with maintaining their current grade of service and/or system coverage. Systems in congested areas cannot get additional channel allocations, forcing them to consider expensive trunking systems or adopt narrowband operation.

Systems in uncongested areas can usually get additional channels by submitting a request to the FCC, but the operators of these systems are concerned that the FCC will institute licensing changes that will affect their future. The intent of refarming is to ease spectrum crowding and congestion. The FCC will achieve this by splitting the current channels, making additional channels available. This is the primary positive effect of refarming and is a broadly supported goal of system operators, equipment manufacturers, and the FCC.

The final FCC refarming R&O eases the transition to the new channel plan by maintaining common channel centers for wideband and narrowband channels. Thus, both wideband and narrowband equipment can be operated on the same frequencies.

The urban PLMR user community is demanding relief from congestion; however, there is no simple solution. The approach selected by the FCC requires users to face potential effects such as future system performance degradation, equipment interoperability problems, single-vendor sourcing because of proprietary technology, equipment availability uncertainty, equipment pricing uncertainty, narrowband transition costs, and operational problems during migration.

These effects are consistent with the results of the transit system survey conducted in Task 2 of this project. Table 16 summarizes these effects for congested and uncongested transit radio systems during the two major time frames of interest. The sections following the table provide detailed descriptions of each specific effect listed in the table.

Congested System Effects

12.5-KHz Dual-Mode Equipment Pricing

The FCC prefers that new systems installed after August 1, 1996, have channel spacing of 12.5 KHz. The implication for the transit operator is increased equipment costs, as opposed to simply replacing wideband analog equipment. Dual-mode or single-mode 12.5-KHz equipment is relatively new, not produced in high quantity, and available from only a few vendors. All these factors lead to higher prices. Transit systems can expect to pay a minimum of 5 to 10 percent more for new 12.5-KHz-capable equipment for the near future.

Narrowband Transition Costs (1995 to 2004)

Transition-related costs will be incurred with the replacement of equipment, upgrades, and new system installations.

TABLE 16 Summary of refarming effects on transit

Time Frame	Congested Systems	Uncongested Systems	Congested & Uncongested
1996 to 2004	• 12.5 KHz Dual-mode Equipment Pricing • Narrowband Transition Costs: - Safe Harbor - Installation - Training - Maintenance	• 25-KHz Single-mode Equipment Pricing	• Equipment Interoperability • Vendor Choice
2005 and Beyond	• 6.25-KHz Dual-mode Equipment Availability • 6.25-KHz Dual-mode Equipment Pricing • Narrowband Transition Costs: - Safe Harbor - Installation - Training - Maintenance • 6.25-KHz Digital System Performance	• 25-KHz Single-mode Product Availability • 25-KHz Single-mode Equipment Pricing	• Equipment Interoperability • Vendor Choice

These additional costs are a consequence of (1) reengineering of base stations because of the new safe harbor requirements, (2) increased installation costs for switching to dual-mode equipment, (3) increased training costs for operation and maintenance of the new equipment, (4) higher maintenance costs because of increased complexity of equipment, and (5) additional maintenance costs because of the need to invest in new maintenance parts and/or test equipment. These additional costs may be incurred twice, first during the 12.5-KHz transition and again when the 6.25-KHz transition becomes necessary. The magnitude of these costs depends on the particular system's size and complexity.

Many transit systems are planning near-term purchases that include new base station equipment and may or may not include making the transition to narrowband. The safe harbor (ERP/HAAT) restrictions imposed in the FCC R& O may require these systems to either reduce transmitter power or relocate their antennas to lower elevations (HAAT). Therefore, all transit systems should analyze their planned base station configurations to ensure compliance with the new ERP/HAAT rules. Reducing transmitter power or lowering the antenna may have no noticeable effect on system performance. On the other hand, it could cause gaps or holes in coverage because of clutter (obstructions or terrain blocking the signal). In the worst case, it may be necessary to reengineer the system, alter antenna patterns, or add base stations (simulcast) to fill the coverage gaps.

The possible effects of meeting the safe harbor requirements are reduced performance (in reliability or channel clarity), service interruptions during system upgrade, added costs for system reengineering (e.g., for additional base stations, antennas, or related equipment to maintain adequate coverage), and additional training for installation and maintenance of new equipment.

6.25-KHz Dual-Mode Equipment Availability

After January 1, 2005, all new PLMR type-accepted equipment must be capable of 6.25-KHz channelization. To date, no vendor offers 6.25-KHz equipment for the traditional PLMR bands. Equipment is under development for use on the band, but only 220-MHz mobiles exist today (no portable handsets). Equipment vendors offer no assurances on the availability, pricing, and performance of 6.25-KHz equipment. This situation is the basis of the user's negative reaction to the initial FCC refarming plan. Vendors, however, are reporting that they will have 6.25-KHz products type-accepted by the 2005 deadline. This uncertainty is a cause of concern for transit systems planning to purchase new systems in the year 2005.

Transit radio system users can be reasonably assured of other effects resulting from a transition to 6.25 KHz, including the following:

- System reengineering and increased number of base stations and related equipment to maintain adequate coverage,
- Increased training costs for installation and maintenance of new equipment,
- Higher prices for new technology, and
- Additional expenditures for spare parts and possibly test equipment.

6.25-KHz Dual-Mode Equipment Pricing

Narrowband 6.25-KHz digital equipment is likely to cost more than wideband digital products to achieve acceptable voice and data performance. The narrowband solutions offered by major equipment vendors are digital in nature with a strong emphasis on trunking. Trunked systems have certain operational advantages but are significantly more expensive than conventional systems. Other vendors probably will follow the product direction of major vendors. Vendors of other advanced narrowband approaches such as linear modulation have no definitive plans to offer products outside the 220-MHz band.

Narrowband Transition Costs (2005 and Beyond)

Transition-related costs beyond the year 2005 will be similar to the transition costs to 12.5 KHz. These additional costs are a consequence of (1) reengineering of base stations because of the new safe harbor requirements, (2) increased installation costs for switching to dual-mode equipment, (3) increased training costs for operation and maintenance of the new equipment, (4) higher maintenance costs because of increased complexity of equipment, and (5) additional maintenance costs because of the need to invest in new maintenance parts and/or test equipment. The magnitude of these costs depends on the particular system's size and complexity. Numerous transit systems may feel forced to make the transition to narrowband, especially if spectrum fees are in place.

The possible effects of meeting the narrowband and safe harbor requirements are reduced performance (in reliability or channel clarity), service interruptions during system upgrade, added costs for system reengineering (e.g., for additional base stations, antennas, or related equipment to maintain adequate coverage), and additional training for installation and maintenance of new equipment.

6.25-KHz Digital System Performance

Transit radio systems in congested areas will feel the greatest need to make the transition to narrowband digital operation, to gain access to additional channels. This transi-

TABLE 17 Digital system performance and its effects

Digital System Performance (versus Analog)	Impact
Increased interference	• Reduced communications reliability, coverage or voice quality
Need for higher transmitter stability	• Higher equipment cost
Reduced voice quality	• Reduced intelligibility and communications reliability
Higher data transmission capability	• Improved performance on CAD systems or telemetry
Greater battery power requirements	• Higher prices for improved batteries • Reduced portable battery life

tion will be accompanied by operational performance and cost effects. These effects, relative to wideband analog systems, are listed in Table 17. The following subsections discuss each of these effects in greater detail.

Increased Interference. The performance of any radio system depends on how much stronger the desired signal is than the undesired noise and interference signals in the same channel. The use of narrower-band channels by all users in a region will reduce the likelihood of co-channel interference but will place more stringent requirements on the receiver filters to reject adjacent-channel interfering signals while not distorting the desired in-channel signal. Adjacent-channel interference will also become more prevalent because the FCC is proposing to eliminate the practice of assigning only alternate channels in a particular geographic area. Regions with a mix of narrowband and wideband equipment may also experience increased interference. A transit operator will no longer be assured that the interfering transmitters operating on the adjacent channels will have sufficient spatial separation from the operator's system.

Need for Higher Transmitter Stability. Transmitter stability also becomes more critical as the channel bandwidth becomes smaller. A transmitter is designed to operate at the assigned center frequency of the channel. Because all transmitters have some drift or instability, the radio cannot effectively use the entire channel without risk of imposing interference in the adjacent channels. Transmitter stability is rated in parts per million (ppm) as a fraction of the transmitter frequency (this problem is more severe at the higher UHF frequencies).

The best crystal oscillators can achieve about 0.0002-percent accuracy, which at 512 MHz is approximately 1 KHz. In a 6.25-KHz channel, a 1-KHz drift means that the radio designer must reserve a larger portion of the already scarce channel for guard band, leaving less available for the actual transmission. As a consequence, the designer must use transmitter components with the highest possible stability. The net effect is increased equipment cost because of the need for higher frequency stability.

Reduced Voice Quality. The voice quality of a narrowband system depends on whether (1) the system is digital or analog, (2) the receiver is operating with a good RF carrier-to-noise (C/N) ratio, and (3) the unit is operating at the border of the service area (i.e., the carrier signal is near the receiver threshold). A digital system employs a coder-decoder (CODEC) to convert the analog signal to a digital representation for transmission and to convert the received digital signal back to an analog voice signal. Digital PLMR systems also employ compression schemes to reduce the amount of data for the digitally coded voice signal. This is accomplished by removing redundant information, modeling the human voice, and applying predictive algorithms.

As a consequence of reducing the amount of data to transmit voice, compression tends to multiply errors. Therefore, these systems also employ some form of forward error compression (FEC). This enables the system to operate in a less than perfect environment, in the presence of signal impairments such as interference, shadowing, or multipath distortion. The specific CODEC, compression, and FEC scheme varies from manufacturer to manufacturer, making it difficult to compare performance of different systems, and results in incompatibility between equipment manufactured by different vendors.

In a well-planned and designed system, the RF-received carrier level will be well above the RF noise level throughout the desired coverage area, providing a C/N ratio over the receiver threshold level. This threshold level is defined as the limiting point at which the voice signal becomes so distorted and noisy that it is no longer intelligible. In an analog FM system, the quality of the voice signal can be quantified by sending a test tone over the radio system and measuring the audio signal to noise and distortion (SINAD). SINAD is related to the RF C/N level. The performance of different FM systems may be compared by measuring the SINAD at various values of C/N.

Because the compression mechanism of a digital system is designed to model the human voice and not audio tones, it is not possible to pass a test tone without adding distortion, and the distortion performance of a system carrying a tone may not indicate the performance of the system carrying human voice. Furthermore, the performance of a particular compression scheme may vary greatly from one speaker to another because of tone, inflection, talking speed, clarity, and volume.

Digital radio system performance is largely defined by two elements: the RF receiver/transmitter (analog and digital) and CODEC/compression/FEC (digital). The relative performance of two competing radio products may be compared using these elements. The CODEC/compression/FEC performance may be defined as the minimum bit-error rate required to achieve the desired subjective mean opinion voice score (or minimum intelligibility), and the RF receiver/transmitter performance may be quantified by the minimum RF C/N required to achieve the necessary bit-error rate.

The above discussion assumes that there is a specific threshold RF C/N below which the system is nonfunctional. In fact, digital systems tend to perform in this manner, in which the voice quality remains good until the system RF C/N reaches a threshold at which the FEC can no longer compensate for the introduced bit errors. At this point, the speech inflection changes randomly, resulting in what has been described as the "Darth Vader effect." There is a very fine line between this point and the point at which the speech is unintelligible and the digital equivalent of "squelch" shuts the receiver off.

Analog systems, which most transit systems use, tend to suffer voice quality degradation gracefully, beginning at higher RF C/N levels than a typical digital system but degrading in smaller degrees as the C/N is reduced. When compared in the same environment, the voice clarity of a digital system is better than the analog system as the user moves into the coverage fringe areas, but as the user moves out even farther from the base, the analog system is still marginally usable when the digital system is no longer functional.

Analysis of the voice quality and coverage performance of the narrowband products offered by the various suppliers is beyond the scope of this project; however, vendors claim that their narrowband digital products provide comparable coverage (for an acceptable voice quality) to their analog wideband PLMR systems and that their analog linear-modulation narrowband radio provides the same coverage as today's wideband FM systems.

Higher Data Transmission Capability. Reducing the bandwidth effectively reduces the data throughput or data speed that can pass over the channel. One would then expect that the new narrowband channels proposed under refarming would reduce the data transmission capability available to the PLMR user. In fact, the analog FM systems were designed to carry voice, and the transmission of data is constrained by the design. Digital systems, on the other hand, are designed to carry data efficiently, because these systems convert the analog voice into a data signal for transmission.

The new applications identified by transit systems will begin to emerge by the year 2005. These applications place heavier data requirements on the radio system. However, the message rate and message length for these new applications are quite low. Digital narrowband systems of 12.5 KHz can reliably carry data at throughputs of 7.2 Kbps or above, whereas most FM systems are capable of only 2.4 Kbps. These new systems are using the latest modulation technologies to achieve this higher throughput. The move from 12.5-KHz channels to 6.25-KHz channels will probably result in a data throughput reduction as opposed to 12.5-KHz digital, although the resultant throughput will still be higher than that provided by the typical analog FM system.

TDMA systems can potentially assign multiple time-division channels on demand to provide greater aggregate data throughput and offer a potential advantage for data transmission over other narrowband technologies. The transition to narrowband channels will probably result in the positive effect of higher data transmission throughput. However, at least initially this will cost more than analog FM.

Greater Battery Power Requirements. Even though the transmitter power will not change in the shift from FM to narrowband, the high-speed digital signal processors required to encode, compress, filter, decode, and decompress the signal require significantly higher power to operate, thereby putting additional demand on the battery of a portable radio. In fact, it has been reported that early public safety users of digital portable radios were not able to get a full shift (8 hours) out of a single battery.

Battery technology is advancing to solve the problem and it is not felt that this will remain a major concern. New battery chemistries, intelligent batteries, and power management intelligence in the radio are all promising to provide longer life from smaller batteries. Added features such as accurate estimation of remaining battery life and elimination of battery memory effects common to NiCd batteries will also be possible; however, that will lead to increased cost per battery, a major recurring maintenance expense for transit systems.

Uncongested System Effects

25-KHz Single-Mode Product Pricing

Transit systems operating in uncongested areas will not feel compelled to adopt narrowband technology, and the FCC R&O permits them to use older type-accepted wideband FM systems indefinitely. Many transit systems considering equipment purchases are staying with analog systems because of their lower price and proven performance. These systems are a mix of conventional and trunked systems.

Changes in pricing of conventional 25-KHz analog wideband equipment may affect these transit systems as the vendors of the older wideband equipment concentrate their production and marketing efforts on the new type-accepted, dual-mode equipment. Dual-mode digital radio equipment carries a 25 to 40 percent price premium over equivalent single-mode analog products. Single-mode wideband products will remain an attractive, "safe" purchase for transit systems with uncongested service areas. Single-mode prices could decline as dual-mode prices decline or single-mode prices could increase as production volume decreases and suppliers discontinue models.

25-KHz Single-Mode Product Availability

Transit systems could be negatively affected by reduced availability of single-mode wideband analog equipment. A need for wideband conventional analog equipment will remain for at least the next 10 years and possibly longer, depending on the narrowband adoption rate. Equipment manufacturers may discontinue production of models with low demand. Fewer manufacturers will offer fewer models, affecting transit systems as they struggle to operate and maintain their wideband systems.

Congested and Uncongested System Effects

Equipment Interoperability

The FCC refarming decision may affect the ability of transit systems to purchase compatible equipment from multiple sources. Transit systems purchasing new narrowband equipment may be locked into purchasing all future expansions and replacement equipment from one vendor, because of the proprietary nature of the equipment. Furthermore, transit systems that wish to communicate with other PLMR users (e.g., other transit systems that share resources, police, fire, or other maintenance personnel) may be able to do so using only the wideband mode of dual-mode radios or could require an additional radio to achieve interoperability.

Vendor Choice

It can be argued that the requirement for narrowband channels will further decrease equipment options, as large vendors introduce refarming-compliant products and smaller vendors wait to determine what the larger vendors are going to offer. The potential effect of this situation is higher equipment prices. The transit industry has limited options to remedy this situation beyond pursuing and adopting an industry standard or using private network services.

Effect Summary

The above effects translate into capital and operational cost increases for transit systems. The extent of the effect depends highly on the specific situation of each transit system. Many of the transit systems surveyed will have replaced all or part of their radio systems by the year 2000. Thus, refarming could have a particularly high effect on those without migration strategies.

Radio systems complying with refarming rules will use technologies that both improve and compromise performance and that present the buyer with difficult-to-compare performance specifications. It is recommended that any transit entity in the market for a new system undertake a thorough investigation to ensure that the new system meets the present and future needs of the transit system.

Hypothetical Effect Examples

The research team performed a hypothetical analysis of the refarming cost effect on transit radio systems. The research team used systems that participated in the survey for this analysis, making assumptions on equipment age and investment decisions. One system is a new, medium-sized, state-of-the-art system; the second is a large, 10-year-old system; and the third is a very small, 7-year-old system. The analysis considered only the incremental cost of refarming. The research team used the transition dates of August 1, 1996, and January 1, 2005.

FCC Rule Assumptions

The FCC has only recently issued initial refarming rules through the R&O (June 15, 1995) and opened a new activity through the FNPRM. The research team assumes no major reconsideration of the R&O will occur—the rules will stand as issued. The research team also assumes the FNPRM will have no effect over the time frames considered. The early stage of this FNPRM and likely delays in a subsequent R&O prohibit the consideration of its eventual effects. The research team also assumes radio services will consolidate before August 1996 and will consist of four services.

Typical Radio Equipment Useful Life Assumptions

Typical useful life for transit radio equipment, as indicated in transit system interviews, is shown in Table 18.

Although transit systems in reality use some of their radio equipment for longer periods than shown, the research team used these life figures in the analysis.

TABLE 18 Radio equipment useful life

Equipment Type	Useful Life
Portable	7 years
Mobile	12 years
Base Station, Repeater	20 Years

Equipment Cost Estimates

Equipment cost estimates are based on information provided by suppliers, assuming the typical features required for transit applications. Assumptions are shown in Table 19. To simplify the analysis, the research team made other key assumptions as follows:

- Narrowband systems of 6.25 KHz will be digital in nature.
- All transit radio systems will make the transition to narrowband as soon as practicable, largely because of concern about high future user fees.
- The high end of portable and mobile prices from Table 19 are used. All figures and product pricing are based on 1995 dollars.
- By the year 2005, dual-mode equipment will remain higher in price than single-mode equipment.
- Dual-mode 6.25-KHz equipment will be available by 2005, but the products will be new and will be produced in relatively low volumes. Thus, a price premium of 25 percent for dual-mode 6.25-KHz equipment over dual-mode 25/12.5-KHz equipment is assumed.
- Dual-mode 6.25-KHz base station equipment is trunked.
- The time base will begin in 1995; thus year 5 is 2000.
- An additional 25 percent over the equipment cost is assumed to cover installation and training costs.

Medium-Sized Multimode Transit System A Example. System A operates a 2-year-old, 450-MHz, 10-channel, digital, trunked system. The channel allocations are exclusive use. Currently, 1,200 mobiles, 139 portables, and 22 base stations are used. Growth in paratransit and light rail vehicles is expected. The system is a 25-KHz trunked system, capable of operating at either 25 or 12.5 KHz.

Short-term effects on System A would be minimal. The 12.5-KHz equipment requirement has no effect, because it is an existing system. In year 10, the research team assumes the system has 1,400 mobiles and 300 portables. All 1,400 mobiles are replaced with dual-mode 6.25-KHz equipment in year 10. In year 12, all portables are replaced with dual-mode 6.25-KHz equipment. The base station equipment must be replaced 8 years early to convert to 6.25-KHz channels. The incremental cost of this 6.25-KHz transition is as follows:

1,400 mobiles × [$1,800 × 0.25]	= $0.63M
300 portables × [$2,300 × 0.25]	= $0.17M
22 base stations × $18,000 × 8/20 years × 1.25	= $0.20M
Installation/training (0.25x factor)	= $0.25M
Total 6.25 KHz transition costs	= $1.25M

Large Rail Transit System B Example. System B has a 10-year-old radio system for its heavy rail operation. It has

TABLE 19 Equipment cost estimates

	Conventional			Trunked	
	Analog	Analog Dual-Mode 25/12.5-KHz	Digital	Analog	Digital
Portables	$1,500-2,000	$1,800-2,300	$2,500-3,000	$2,000-2,500	$3,000-3,500
Mobiles	$1,000-1,500	$1,200-1,800	$2,000-2,500	$1,500-2,000	$2,500-3,000
Base Station	$10,000	$12,000	$12,000	$15,000	$18,000

13 base stations, 1,800 mobiles, and 2,000 portables provided by two vendors. The research team assumes the mobiles have 2 years' useful life remaining and no dual-mode equipment is in use. The research team assumes dual-mode equipment from its two vendors are not compatible. System B will be forced to select one vendor for future purchases.

Immediate effects on System B would be minimal, assuming the safe harbor ERP/HAAT requirements are met. In year 2, mobiles must be replaced. In year 4, portables and base station equipment must be replaced to implement 12.5-KHz operation. The research team assumes 50 percent of a new digital base station cost is required to upgrade. Incremental 12.5-KHz transition costs are as follows:

1,800 mobiles × $300	= $0.54M
2,000 portables × $300	= $0.60M
13 base stations × $9,000	= $0.12M
Installation/training (25% of equipment costs)	= $0.32M
Total 12.5-KHz transition costs	= $1.58M

In years 11 through 14, all 1,800 mobiles, 2,000 portables, and 13 base stations will be replaced with 6.25-KHz equipment. The incremental costs for the 6.25 KHz transition are as follows:

1,800 mobiles × [$1,800 × 0.25]	= $0.81M
2,000 portables × [$2,300 × 0.25]	= $1.15M
13 base stations × [$18,000 × 0.25]	= $0.59M
Installation/training (25% of equipment costs)	= $0.64M
Total 6.25-KHz transition costs	= $3.19M
Total 12.5-KHz transition costs	= $1.58M
Total refarming-related costs	= $4.74M

Small Taxi Transit System C Example. Taxi transit system C has a 7-year-old, 150-MHz, analog system with 2 base stations and 38 mobiles. In year 5, mobiles will be replaced and base stations will be upgraded to dual-mode 12.5-KHz equipment. Transition costs are as follows:

38 mobiles × $300	= $11,400
2 base stations × $9,000	= $18,000
Installation/training (25% of equipment costs)	= $ 7,350
Total 12.5-KHz transition costs	= $36,750

In year 17, the entire system must be upgraded to 6.25-KHz equipment (dual mode or single mode). Because the base stations will be replaced regardless, incremental costs of moving from analog equipment to 6.25-KHz equipment are included.

38 mobiles × [$1,800 × 0.25]	= $17,100
2 base stations × [$6,000 + $18,000 × 0.25]	= $12,000
Installation/training (25% of equipment costs)	= $ 7,275
Total 6.25-KHz transition costs	= $36,375
Total 12.5-KHz transition costs	= $36,750
Total refarming-related costs	= $73,125

Table 20 summarizes the cost effect for the three systems. These three transit systems face varying effects, roughly related to their size. The cost magnitude represents a significant percentage of typical new system costs. Pending fees or auction costs will add to the cost effect. Newer systems appear most susceptible to high relative refarming costs because of early retirement of equipment.

POSSIBLE RESPONSES TO REFARMING

Transit systems must carefully consider the near- and long-term effects of refarming when planning changes to their radio systems. Failure to do so may result in unneeded costs, early obsolescence, and reduced quality of service.

TABLE 20 Transition cost effect summary

System	12.5-KHz Transition Costs	6.25-KHz Transition Costs	Total Transition Costs
System A - Medium, 2-year-old Multimode	-	$1,140,000	$1,230,000
System B - Large, 10-year-old Rail	$1,570,000	$3,170,000	$4,740,000
System C - Small, 7-year-old Taxi	$36,750	$47,625	$84,375

In this section, the research team describes feasible action plans that the transit industry may employ.

Migrate to Narrowband

A transit system, particularly in congested areas, could accept the FCC incentive for migration to narrowband. The additional costs may cause budgetary problems, but rapid conversion that relieves congestion problems may be justifiable. A migration to 12.5 KHz is relatively low risk, because equipment is available.

Defer Major Investments

Deferring major investments in radio equipment is a conservative action that transit operators could consider. This deferral would continue until (1) narrowband equipment availability is clearly understood and (2) the FNPRM activity is resolved. It is hoped that the latter will define how newly created channels will be allocated, enabling transit operators to efficiently resolve congestion problems. The period for this investment deferral could be 1 to 3 years. That assumes the current radio system will suffice in the interim.

The uncertainty surrounding the FCC refarming increases the risk of near-term investments in radio systems. Transit organizations may waste capital resources if communications systems do not comply with the new rules or if they produce high spectrum-use fees. Many medium and small transit operators do not have a good understanding of the FCC activities, placing them at greater risk of unwise decisions.

The costs of deferring investments may include continued inefficient daily operations and high system maintenance costs. Many transit systems interviewed during the survey portion of this project cited obsolescence as the primary reason for planned radio system investments. Long investment deferral may decrease the availability of replacement parts and increase reliability problems. Systems experiencing operational problems, such as congestion and poor coverage, because of aging equipment, will continue to do so. Those operators electing to defer will have to tolerate these problems for the deferral period.

Move to the 800/900-MHz Frequency Band

Moving into the 800/900-MHz frequency band is another possible action for transit. This alternative applies only to those areas where channel allocations are available. Some congested areas, such as New York City, have no available 800- or 900-MHz channels.

Equipment for use in these bands is available from the traditional PLMR vendors. One caution is that 800/900-MHz channels are wideband (25 KHz). These bands could eventually also be refarmed, although they are relatively new bands and the FCC has not announced any plans for refarming this spectrum.

Transition to higher frequencies requires higher investment costs than conventional PLMR equipment. Systems of 800/900 MHz are typically trunked, which costs at least 50 percent more than conventional systems. The higher propagation attenuation characteristics at 800/900 MHz may also necessitate system redesign and relocation or addition of base station sites. The transition may not be as simple as replacing the old base stations.

Operational issues relate to the general advantages and disadvantages of higher frequency trunked systems. Congestion will be minimized with a sufficient number of channels and privacy is possible. Signal penetration through certain barriers such as glass is good, which is helpful in a transit application.

Use Public or Shared Private Networks (Nonsafety-Critical Functions)

Transit systems could use public or private WAN services for nonsafety critical functions of the operation. Procuring radio network services could allow deferral or elimination of a radio system capital investment. Many transit systems in the survey use cellular voice and basic paging WAN services. Numerous public network options exist, as shown in Table 21.

Private network operators are faced with justifying large investments in their private networks. Some of these private network operators are offering excess capacity to other users with similar needs to defray their investment. Utilities are one example of this type of network operator. Transit systems can investigate private network operators in their service areas and assess the potential for sharing the radio infrastructure.

The cost of using public or shared private WAN services depends on user loading and competitive pricing pressures. Transit systems could purchase handhelds and mobiles from

TABLE 21 Public WAN service options

	Voice	Data
Existing	Cellular SMR Mobile Satellite	ARDIS RAM Advanced Paging Mobile Satellite
Emerging	PCS Enhanced SMR	CDPD Data PCS

multiple WAN equipment vendors. This user equipment should be relatively inexpensive because of industry standards and large numbers of users. Base stations and antenna sites will not be required. Capital and maintenance costs will be lower than purchasing a private system, but operational costs will probably be higher.

The cost savings and reduction in risk may off set the loss of operational autonomy. Smaller transit systems, including paratransit functions, may find this approach attractive. Refarming is not an issue for WAN radio services, because most operate above 800 MHz.

The research team cannot fully define the operational opportunities for, and implications of, using public or shared private WAN services. The use of WAN radio services warrants investigation. This option was not available when most radio systems were originally installed. The costs for WAN user terminal equipment and network access continue to decline, increasing the attractiveness of this option.

Retain Channel Allocations in the PLMR Bands

Transit systems should retain rights to any channel allocations in the PLMR bands. Regardless of other actions, this will allow the flexibility to optimize system performance for future applications. Often this approach is not possible if the FCC issues new channels contingent on the release of old channel allocations. Retaining a few contiguous channels would permit use of high data rate applications in the future.

Assess Compliance with Safe Harbor Table

Transit systems should determine their compliance with the safe harbor ERP/HAAT table for ongoing base station installations. The analysis required to verify compliance is straightforward. The analysis can provide either a direct comparison of each base station configuration with the table per the final ruling or a system-level coverage contour simulation by an equipment vendor or an independent consultant to document coverage versus service area compliance.

Embrace or Develop an Open Standard

The transit industry could benefit by using or developing an open communications standard. Transit systems have limited equipment choices and limited competition among vendors. The refarming activity threatens to worsen this situation. A purchase commitment in the near term may lock a transit system into a particular vendor's equipment.

The transit industry could explore existing communications standards that may be applicable to transit operations. This exploration should include U.S. and international standards. The APCO Project 25 standard is one option to be explored. APCO 25 is a new, unproven standard, but APCO has needs that may match well with transit needs. Equipment manufacturers have offered to license their technical approach, as an open standard, at low cost to other vendors. European standardization efforts have increased and may be an option for U.S. transit systems.

Alternatively, transit could pursue a standardization effort, in a manner similar to the APCO 25 effort, to define a standard that fully meets its needs and ensures competition. This would be a major undertaking requiring significant time and funding. APCO, as a large user community, felt justified in its investment, and transit could conceivably also justify an investment of this sort.

Further Research Options

An Effort to Understand the Effect of the FNPRM

TRB Project C-5 was completed on the basis of rules issued in June 1995; however, the FCC did not resolve all the pertinent issues and opened a new proceeding in the form of an FNPRM. This FNPRM revealed the possibility of fees for or auction of channel allocations for all users, excluding public safety. The transit industry would benefit from an effort similar to Project C-5 to fully understand the effects and possible responses. The effort would use knowledge gained in Project C-5 on transit usage and plans for radio communications.

In-Depth Technical Review of APCO 25 Applicability to Transit

The APCO 25 digital radio communications standard has been adopted by the public safety community. It represents a major investment in time and effort to satisfy user needs for interoperability and cost savings through competition. The AAR may eventually consider adopting APCO 25 as a railroad standard. The transit industry should determine if APCO 25 satisfies its needs and if adoption as a standard is useful and appropriate.

In-Depth Technical Review of Channel Access Methods

TDMA, FDMA, and linear modulation (LM) are three major channel access methods vying for use in the PLMR bands. A transit operation choosing one of these methods is "locked" into certain vendors and capability. TDMA, FDMA, and LM equipment is not interoperable. A technical review would be useful to contrast TDMA and FDMA, taking transit operations and FCC refarming into account.

Effect Analysis on Medium and Small Transit Operations

The scope of TRB Project C-5 did not allow detailed analysis of refarming's effects, with definitive costs and operational effect. Each transit operation has unique needs and constraints that limit the benefit of broad analysis. Most large transit operations can hire consultants to perform a detailed effect analysis, but the medium and small operations have limited funds. A valuable benchmark could be developed by selecting one or more medium to small transit operation and analyzing in detail the required investment as a result of refarming rules. The cost of this effort might be shared by several transit operations or by transit associations.

Assessment of Wireless Public Data Networks for Transit Use

Wireless public data networks could enable transit operations to defer or eliminate large capital investments in radio system infrastructure. Many transit systems use cellular phones and paging services. These wireless services did not exist when transit systems originally invested in their own radio systems. More extensive use of these and other networks may be useful to transit systems. An in-depth assessment of availability, capability, pricing, benefits, and risks of wireless WAN public data networks, such as paging and cellular CDPD would be informative to transit systems.

Assessment of Shared Investments in Radio Infrastructure

As local and federal funding becomes more scarce, transit operations must prove that they have explored all reasonable options before making new capital investments. Nonsafety critical communications for maintenance, yard, and some supervisory functions could share an infrastructure with other large private radio system operators. This approach could be useful in areas with limited frequency availability or for transit systems with limited capital funding. The potential benefits and risks of sharing systems and investments with utilities and other large users would be assessed to help transit systems meet this challenge.

Electric, gas, and water utilities are also facing cost and competitive pressures, causing them to consider other business and investment options. One utility in Georgia is actively marketing its recently purchased radio network to state police, government agencies, and transit systems.

Assessment of Industry Desire for APCO 25-like Standards Development

The transit industry could benefit from a communications standard in the manner that APCO 25 benefits public safety radio system users. A standards development effort may be useful if APCO 25 itself does not meet the needs of the transit industry. An effort that identifies the approach, outcome, investment, benefits, and risks of standards development and surveys transit industry receptiveness to the concept, may be valuable.

Assessment of International Transit Experience with Narrowband

European and Asian transit operations employ narrowband technology to varying degrees. Asian countries are making large investments in infrastructure, including transit systems. Although European and Asian frequency allocations differ somewhat from those in the United States, the Europeans' and Asians' experience with narrowband technology can provide useful information to the U.S. transit industry. A review of the usage, performance, availability, and pricing of equipment would aid in addressing the refarming challenge.

APPENDIX A

TRB QUESTIONNAIRE

TRB Questionnaire

System Identification

Name of Transit Organization:
Type of Transportation (Rail, Bus, etc):
Describe size of system
- number of vehicles by type
- geographical coverage
- number of employees using radios
- percentage surface vs. subsurface

Location:
Person(s) Interviewed:
What is interviewee's responsibility relative to the communications system?

Phone
FAX:

Radio System

Type of system (trunked, simulcast, single channel, multi-channel, community repeater, mobile-to-mobile, mobile-to-dispatch, other):

Band(s) of operation:
Number of channels in each band:

	Mobiles	Portables	Base stations
Quantity			
Output Power			
HAAT			
ERP			
Number of channels			
Manufacturer			

Communications Requirements:

How is the radio system(s) used? (e.g. dispatch, maintenance, scheduling)

...is it for voice only?
...data?
...telemetry?

What is the extent of deployment of the radio system? (e.g. on every bus, train, supervisors cars)

Is the performance of the transportation system dependent upon the radios (i.e. will the system stop or be seriously degraded if the radios cease working)?

Is the protection of life and property dependent upon the radio system ? (explain)

What is the area of radio coverage required by your system?

is this the same as the geographical responsibility of the transit system?

What is the distance to a neighboring radio system on the same channel?

What ERP levels exist on the system (min, max, typical)

Are there other required characteristics of your radio system? (e.g. portability, ANI, mobile repeater, talk-around, coded squelch, telephone interconnect, extremes of weather, harsh environment, noisy environment, etc.)

...is portability a requirement, if so, what is the acceptable battery life you can tolerate?

Spectrum Allocations

How many channels are you presently using?

Are the channels shared or exclusive use?
Have you requested additional channels in the past?
...what was the outcome (refused, pending, other)?

...if pending, do you expect to receive the allocation?
Are your requirements increasing? please elaborate

Have you plans to request additional channels in the future?
...how many and why?

Capital Investment

How much have you invested in your radio system?
...in the last 5 yrs?
...past 10 yrs?
What is your forecast requirements for next 2 yrs?
...5 yrs
...10 yrs.
How are the purchases financed?

if financed through bonds, when will existing bonds be retired and what is typical bonding period?

Problems Experienced

Have you experienced problems with your radio system because of heavy usage(congestion, contention, interference, etc.)? Please explain

How do these problems impact your business?
...insufficient coverage?
...how often do these problems occur?
...what percentage of your radio calls are affected?
...only during certain times of day?
...certain days?
...explain:

Please discuss any interference problems you have experienced with your neighbors or other radio systems (voice, data or telemetry)

Do you mix voice data or telemetry on the same channel? why or why not?

What problems with mixing have you encountered, if any? How have/will you resolve these problems (new channel requests, new equipment, etc)

If trunking or other advanced features are used, why and do they in fact achieve the desired effect?

Do these features inhibit desired usage or operation in any way?

Advanced Technology

Are you utilizing other forms of wireless communications such as Cellular Telephone, SMR, CB?
...do you plan to?

Explain?

Have you, or do you have plans to invest in narrow band or digital radios?
Have you, or do you have plans to invest in other radio dependent applications (e.g. vehicle location, other)?

Refarming Plans

Are you aware of the FCC refarming activity?

What plans, if any, do you have to deal with the FCC refarming issue?

...have you an estimate of the cost impact?

APPENDIX B

TRB MOBILE RADIO USAGE SURVEY

Transportation Research Board Mobile Radio Usage Survey

Introduction

Arthur D. Little, Inc. (ADL) is under contract with the Transportation Research Board of the National Research Council, which is the principal operating agency of the National Academy of Sciences and the National Academy of Engineering. Our work is sponsored by the Federal Transit Administration (FTA) and supported by the American Public Transit Association (APTA).

ADL's charter under this contract is to report on the impact impending FCC private land mobile radio rule changes will have on the future development and operations of transit systems. To do this we must fully understand the present state of, and future plans for, radio communications within these systems. To achieve this understanding we will survey a representative sample of transit agencies.

Survey Participants

A representative sample of transit agencies was chosen to achieve a balanced set of responses, based upon:

The type of transit system
The location of the system relative to population density
The size of the system
The likely requirements for radio usage both present and future.

Your organization has been selected as an important survey participant. All information obtained will be used to develop an understanding of how the FCC's proposed rule changes will impact your:

Ability to obtain appropriate spectrum allocations
Budget requirements for new equipment mandated through new technical requirements.
Ability to achieve sufficient radio coverage with your present or future systems.

A summary of this survey and our analysis will be reported at the 1995 APTA annual meeting in San Antonio, Texas.

The Questionnaire

We will contact you via telephone at the agreed upon time, to conduct a brief discussion. Each question we will ask relates to an aspect of the radio communications that is subject to impact by the FCC's proposed rule changes. For example, questions relating to recent or planned capital expenditures are intended to quantify the budgetary impact that will result if the FCC mandates new technical specifications forcing early retirement of equipment. We are not attempting to question the validity or soundness of your capital expenditures.

We also recognize that transit systems requirements for wireless communications are on the rise and will continue to rise with the advent of advanced applications and new enabling technologies. We will attempt to incorporate these new requirements in our analysis to the extent that transit systems operators will share with us their vision of the future.

Question Topics

System Description
- size, location and description of your transit system

Radio System Description
- Description of your radio system
- Numbers of radios of each type, channels, operating bands, output power, radio manufacturers, etc.

Communications Requirements
- How is your radio system used?
- What is the geographic coverage?
- Are there any special performance requirements, features?

Spectrum Allocations
- Are your channels shared with other users?
- What are your future channel requirements?

Capital Investment
- How much have you invested in your radio system?
- What have you budgeted and when do you plan for additional investments?

Problems Experienced
- Do you experience problems with your existing radio system? Please be specific.

Advanced Technology
- Are you using, or plan to use, advanced radio technologies (cellular, vehicle location, digital, trunking, etc.) for new applications or improved performance? Please be specific.

Refarming Understanding and Plans
- Have you made specific plans to deal with FCC Refarming?
- Can you provide an estimated cost?

Please be prepared to discuss each of the above topics. Your assistance is greatly appreciated and your response to this inquiry will demonstrate the commitment of the transit industry to improve public service and assure the FCC that supporting the current and projected communications requirements of the transit industry is in the public interest.

APPENDIX C

PRIVATE LAND MOBILE RADIO AND FCC TERMINOLOGY

Private Land Mobile Radio and FCC Terminology

PLMR is segregated into frequency bands, each consisting of a number of channels. Each simplex channel can support voice communications (or other types of communications) for one or a number of users, on a time-share basis in which one person speaks at a time. The channel is effectively a party line.

The frequency of the band affects the performance characteristics of the signal transmitted. In general, lower frequency radio has a relatively greater geographical range. Low-frequency VHF radio signals are known to travel for thousands of miles under certain atmospheric conditions. As a consequence, radio systems operating on the lower frequency bands are more prone to interference from radio systems operating in neighboring areas.

A PLMR license authorizes the license holder to operate a radio system on a particular channel, within a particular band, in a specific geographic area. A frequency coordinator assigns the radio channel, usually on the basis of the loading required for the radio system and the loading that already exists in the applicable channels, in the specific geographic area.

Notice of Proposed Rule Making (NPRM)—A public notice issued by the FCC that contains the form and content of a rule change that has been proposed. It is issued to solicit public comment and alternative proposals.

Effective Radiated Power (ERP)—A measurement of the power radiated from the antenna, based upon the combination of the transmitter output power and the antenna gain

Antenna Gain—A measure of the increase in signal strength resulting from the antenna focusing or confining the radio signal in a desired direction

Height Above Average Terrain (HAAT)—A term used to describe the position (measured in feet) of the antenna relative to the average terrain in the immediate vicinity of the antenna tower

Markets—Geographical areas that represent population centers, usually corresponding to a metropolitan area

Report and Order (R&O)—The final approved rule change instituted by the FCC

Notice of Inquiry (NOI)—A publicly issued request from the FCC for input, comments, and suggestions regarding policy.

Code of Federal Regulations (CFR)—Contains the parts that relate to PLMR regulations and other rules

Part 90—The part of the CFR in effect that relates to PLMR

Part 88—The new rules proposed by the "refarming" NPRM to replace Part 90

Service—PLMR channels are divided into services. Each service groups the licensees by the service that the user performs. Examples of services are Police Radio Service, Fire Radio Service, Forestry Radio Service, and Railroad Radio Service.

Channel Loading—Refers to the degree of use of a channel, measured by the number of mobiles assigned to the channel

Frequency Reuse—Implies limiting the power of the transmitted signal to confine the signal to the desired geographical area and constrain the signal spill-over in neighboring geographical areas where the licensee does not need coverage. The net result is that the same radio channel can be reused by a different licensee in the neighboring geographical area.

Primary/Secondary Licensees—A primary licensee takes preference over a secondary licensee. If the secondary user causes interference to a primary user, it must discontinue operation.

Offset Channel—Frequency allocations between normal channels used for low-power secondary services